T0339432

Managing Supply Chains on the Silk Road

Strategy, Performance, and Risk

Managing Supply Chains on the Silk Road

Strategy, Performance, and Risk

Çağrı Haksöz
Sridhar Seshadri
Ananth V. Iyer

CRC Press
Taylor & Francis Group
Boca Raton London New York

CRC Press is an imprint of the
Taylor & Francis Group, an **informa** business

Cover credit: Book Cover Art by Erkan Kusku.

CRC Press
Taylor & Francis Group
6000 Broken Sound Parkway NW, Suite 300
Boca Raton, FL 33487-2742

First issued in paperback 2018

ISBN-13: 978-1-4398-6720-4 (hbk)
ISBN-13: 978-1-138-37454-6 (pbk)

Library of Congress Cataloging-in-Publication Data

Managing supply chains on the Silk Road : strategy, performance, and risk / editors, Çagri Haksöz, Sridhar Seshadri, Ananth V. Iyer.
    p. cm.
Includes bibliographical references and index.
ISBN 978-1-4398-6720-4 (hardcover : alk. paper)
    1. Business logistics. 2. Silk Road. 3. Trade routes--Asia. I. Haksöz, Çagri. II. Seshadri, Sridhar. III. Iyer, Ananth V.

HD38.5.M3624 2012
658.7--dc23   2011043273

Visit the Taylor & Francis Web site at
http://www.taylorandfrancis.com

and the CRC Press Web site at
http://www.crcpress.com

In memory of Sakıp Sabancı (1933–2004)

To my family (Mehmet, Necla, Burcu) and to unknown souls whose efforts and hard work paved the way for today's global supply chains...

Çağrı Haksöz

To my family...

Sridhar Seshadri

To my family...

Ananth V. Iyer

Contents

Foreword

For almost 3000 years, the Silk Road has enabled merchants and firms to trade products ranging from silk to spices, horses to medicine, and gemstones to textile along the trade routes across the Asian continent. Despite geographical dispersion and language and cultural differences, the Silk Road has connected East, South, and Western Asia with the Mediterranean world, as well as with North, East, and Northeast Africa and Europe. To a great extent, the trading activities along the Silk Road lay the foundation of today's global supply chain operations. Specifically, today's global supply chain operations resemble the trading activities that took place for centuries along the Silk Road. The similarities include (1) multiple routes—different silk roads were developed over the centuries to improve logistics efficiencies; (2) multimodal transportation systems—both land and sea transportation systems have facilitated trade along the Silk Road for centuries; (3) logistics hubs—due to the Silk Road extending for over 4000 miles, goods were transported by a series of agents on varying routes and were traded in different logistics hubs located in various oasis towns; (4) business contracts—due to different cultures and judicial structures, traders along the Silk Road have developed different contracts to secure payments and insurance schemes to hedge against various types of risk events; and (5) supply chain risk management—due to potential natural and man-made disasters such as robberies along the Silk Road, traders have developed different types of risk mitigation mechanisms to reduce the likelihood and the impact associated with different risk factors.

Recognizing the Silk Road trading operations plays a critical role in the development of today's global supply chain operations. Haksöz, Seshadri, and Iyer have brought together 28 scholars and business executives from different continents to share their perspectives about past and present trading activities along the Silk Road. These chapters provide a glimpse into the past and describe the new surge of activities in Hungary, Turkey, Lebanon, Israel, Iran, Pakistan, India, Azerbaijan, Uzbekistan, Turkmenistan, and China. Although the focus of this book is on supply chain operations, it examines a wide range of issues arising from a multicultural perspective. It also provides clear insights of the past and the present that will help academics and practitioners to gain a better understanding of the future.

Christopher Tang
University Distinguished Professor
Edward W. Carter Chair in Business Administration
Anderson School of Management
University of California, Los Angeles
Los Angeles, California

Preface

This book has been written by 28 scholars and business executives from across the globe. In this book, we examine supply chain management along the Silk Road from three perspectives: strategy, performance, and risk. Starting with a summary of supply chains along the historic Silk Road from ancient to postmedieval times, we provide descriptions of interesting supply chain practices and insights from regions along the historical Silk Road that played, and continue to play, an important role in global trade.

Our main strategic insight is that the supply chain concept and its related practices that appeared centuries ago along the Silk Road became the foundation for today's global supply chains. Intercultural and interborder operations took new shape in the twenty-first century, yet intermingling of multicultural countries to produce value for end users appeared a long time ago on the Silk Road. We observe that many supply chain practices were originally developed and used centuries ago under somewhat different forms and structures. In this prologue, we identify such practices where possible and provide links between the past and the present. It needs to be noted that today the historical Silk Road itself might not be the actual connecting route since global logistics have changed dramatically. Nevertheless, it is surprising that countries along the Silk Road are still thriving and remain part of the fast-changing world economy. This probably bears testimony to the connection between current economic activity and historical trends.

The book is organized into three parts, Supply Chain Strategy on the Silk Road, Supply Chain Performance on the Silk Road, and Supply Chain Risk on the Silk Road, which present various insights and cases on the *strategy–performance–risk* triangle along the Silk Road supply chains.

In Part I (Chapters 1 through 6), we provide insights on the strategic aspects of supply chain management along the Silk Road, focusing on the historical Silk Road, Central Asia, India, and Israel. In this part, the connection between the historical Silk Road supply chains and today's modern supply chains is made. Later, we focus on how winning supply chain strategies could be constructed in today's Silk Road regions. Turkic countries in Central Asia such as Uzbekistan and Turkmenistan need a different supply chain strategy than India and Israel. Idiosyncrasies of the countries, firms, and their markets play a pivotal role in this process.

In Part II (Chapters 7 through 12), we present various interesting case studies on supply chain performance management in historical Silk Road countries such as China, Turkey, Pakistan, and Hungary. In this part, invaluable case studies on third-party logistics (3PL) management in Turkey, buyer–supplier relationships between European and Chinese firms, significance of the original design manufacturers in global supply chains, milk supply chains in Pakistan, and a potential railroad project between China and Hungary, which aims at building a new Silk Road, are presented.

Lastly, in Part III (Chapters 13 through 16), we focus on various supply chain risks and different methods to manage them effectively. Risks affected the supply chain players' decisions on the historical Silk Road. They continue to impact the design and operations of supply chains along the Silk Road. Strategic risk taking and managing have become inevitable for designing today's Silk Road supply chains. Connectivity, leanness imperative, fickle customer expectations, and volatile world markets create interdependent and unknown risks in today's supply chains. Insights gained through case studies from Gaza, Lebanon, the Asian tsunami in 2004, as well as Turkey and Iran are presented in this part. We present abstracts of the chapters in the following and invite the reader to journey along the Silk Road with us.

In Chapter 1, Çağrı Haksöz and Damla Durak Uşar take a historical snapshot of Silk Road supply chains from ancient to postmedieval times until the nineteenth century. The authors provide factual details on historical Silk Road supply chains that traversed a vast geographic region from China to the Balkans and Europe for a very long period in human history. It is interesting to note that the current supply chain activities and problems existed centuries ago along the Silk Road. In the days past, innovative solutions were devised in different parts of the Silk Road, examples of which are provided in this chapter. Characteristics of the Silk Road trade, players of the Silk Road supply chains, products and services offered, procurement and sales management, lead times, careers of the Silk Road, risk management, role of governments, and managing interborder and intercultural differences are the focus points of this chapter.

In Chapter 2, Ananth V. Iyer and Andrea Lenterna present the case of Indian artisans and their interesting connections with Italian designers. In India, the manufacturing of textiles has a long history. The early Indians designed and produced fabric that was traded on the original Silk Road; currently they carry out a portion of the designing and all of the producing. Indian designers today collaborate with their Italian counterparts to create the latest fashion trends. The case of Arfa illustrates this relationship very well. Italian designers coordinate through Arfa with Indian manufacturers to design the fabric, and Indian manufacturers have the clothes stitched? Arfa maintains the relationship between the two parties. In the ever-changing fashion world, they do business without contracts—partnering with someone you can trust is very important and Arfa helps facilitate the process. In manufacturing, if the processes can be broken down into subprocesses and if each process can be handled by specialized persons, costs will eventually come down. This is how the networks of the Indian fashion industry operate with intermediaries like Arfa. They are coordinated by the intermediary and therefore can either focus on designing or producing the products. This chapter is a reflection of the past into the future. In the past, artisans along the Silk Road were also in high demand in order to produce valuable products and luxury items such as silk cloths and embroidered garments. Today, in many locations, though they are reduced in numbers, a new wave of collaborative supply chain structures enhance their value as showcased in this chapter.

In Chapter 3, Ercan Korkut presents interesting managerial insights into Central Asian supply chains gathered through years of experience. A major part of the historic Silk Road was in Central Asia. Similar to the Silk Road supply chains of the past, high transaction cost still plays an important role and is mostly due to the fact that the players are from different countries and thus from different cultural backgrounds and, therefore, have developed different laws and regulations. One country that has kept the old traditions of trade is Uzbekistan, which still requires a signed contract between buyer and seller. This greatly increases the transaction costs for international trade of intangibles like software downloads. The economic profitability of trade along the route has been constant through the centuries despite these high transaction costs. Some countries along the Silk Road have yet to fully adapt to the new technologies. For example, Russia and the ex-Soviet countries still require a CD to be included in transactions that take place completely online. A handful of countries along the Silk Road are under trade embargoes, so any trade of components to these countries is heavily regulated and therefore must be monitored. Customs duties are imposed by individual countries to protect their fledgling economies. Trading in the countries along the Silk Road can be difficult due to the different trade laws and customs along the route. The best way to effectively trade among these countries is to establish local hubs while still maintaining a global focus. Korkut concludes that collaborating with parties who have a foothold in the local trade, and integrating them into the global organization, will best accomplish trading goals. This practice, having local hubs along the Silk Road, was also used for centuries in historic Silk Road supply chains in order to recognize problems on the spot and manage them effectively.

In Chapter 4, Deniz Tura provides an interesting analysis of bazaars in Azerbaijan, Uzbekistan, and Turkmenistan that played active roles in the historical Silk Road. There is an emerging class of entrepreneurs that has developed since the fall of the Soviet Union in these countries. These entrepreneurs have formed their own type of marketplace to buy and sell wares called a "bazaar." The post-Soviet governments are still in their nascent stages and adjusting to the idea of free market enterprise; they have not developed the laws and regulations to support the fledgling class of small businesses that are developing in their country. As such, they have been forced to develop their own informal systems for doing business. For example, in all three countries, they have developed trust networks to make trade easier. A trust network is a system based on family, ethnic, tribal, or religious ties that aims to smooth business transactions by having a bond with the people you do business with. These trust networks are involved in nearly every aspect of doing business, from supplier credit to importing. Another interesting feature of the informal trade practices that developed is financing. Traditional financing did not meet the needs of the entrepreneurs and thus a different method of financing had to be developed. In Azerbaijan and Uzbekistan, small, informal financial institutions have been created solely to service the needs of small businesses. These institutions can transfer money anywhere in the world, and they have a corresponding money shop that can make small loans to businessmen. Turkmenistan still does not have government cooperation to engage in these kinds of innovations, but burgeoning trade can be expected to find innovative solutions that are remarkably similar to ones employed in the past. The concept of bazaars was instrumental in trading a variety of goods from different regions in historical Silk Road supply chains. It is interesting to observe that this type of exchange structure still operates in many countries along the Silk Road.

In Chapter 5, Ehud Menipaz examines how Israel, a player on the Silk Road for quite some time, competes in today's global supply chains. The Mediterranean basin where Israel is located has been a key part of the route for East–West trade in the past, and it has evolved to remain important in the new Silk Road. Israel has a lack of natural resources and so must produce something else if it wishes to compete in the current economic market. With innovation and a strong backing from the government, Israel has transformed its economy into a very important link in the supply chain of the global economy. Also, the culture in Israel has a lot to do with its recent economic success. The people of Israel place a high social value on succeeding in business. Successful entrepreneurs are looked at as celebrities. The Israeli government has nurtured specific sectors in its economy like technology, medical devices, and R&D and has managed to make them competitive in the global economy by providing funds to the rising industries until they grow strong enough to compete on their own. The government then allows these fields to be privatized. The Israeli government also invests heavily in education with the belief that an educated workforce will help grow the economy much faster.

In Chapter 6, Janat Shah and Debabrata Ghosh bring their considerable experience teaching and doing research related to supply chain management in India to describe the challenges faced by firms doing business in India. India, an important country in the old and the new Silk Road, like China, has reversed centuries of trend of being a supplier and has become an attractive market. However, doing business in India is not easy. Problems abound in every aspect of supply chain management, from location, logistics, and distribution channel design and management to retailing and use of the right technology. Shah and Ghosh suggest that successful firms will have to innovate to meet these challenges. Examples of many different companies who have innovated in different areas, including product design, business models, in-bound logistics, distribution, retailing, and technology infrastructure solutions, are provided. The chapter provides many interesting areas for further research and also attests to the profit potential to be had even in challenging conditions in this fast-growing market on the Silk Road.

In Chapter 7, Murat Kaya and Çağrı Haksöz present the case of a Turkish 3PL firm based in İstanbul, Turkey. Turkey is located in the hub between Europe, Asia, and the Middle East and their influence extends across all borders. Turkey was a major part of the historical Silk Road with an economy devoted to serving merchants who traveled through their country on the trade route. Now, Turkey is attempting to become a major producer of goods along the route. Turkey's economy has been growing at a pace that is matched only by the BRIC countries. It is investing heavily in infrastructure and attempting to become a hub of trade in the region. With the infrastructure growing and with exports and trade rising so rapidly, there is a growing need for efficient logistics providers to move goods around the country. There are many competing firms vying for a piece of the lucrative trade business, but one firm stands out—Borusan Lojistik. They have become one of the largest logistics firms in Turkey by providing their customers value-added services. They achieved a growth of 35% annually from 2000 to 2007 by implementing their six-sigma methodology. The authors describe how Borusan Lojistik effectively manages its 3PL business and presents valuable insights. The concept of 3PL is not new. In historical Silk Road supply chains, various agents and intermediaries played the role of 3PL firms. In the maritime trade, companies such as English East India and Dutch East India operated integrated distribution chains along the Silk Road. Moreover, we should note that the concession system used in the sixteenth century paved the path for the 3PL industry.

In Chapter 8, Gürdal Ertek provides a specific example of a beneficial supply chain strategy of cross-docking for a 3PL firm located in İstanbul, Turkey—one of the main destinations for the historical Silk Road—which served as a hub for the exchange of goods on the route. The historical Silk Road used a complex system of cross-docking where goods were transferred from hub to hub exchanging hands along the way. Moreover, goods were transported via multiple modes such as yaks and camels depending on the topography and the riskiness of the regions. This cross-docking facilitated cost-efficient trade of goods along the road. This contrasts with the popular view that one set of merchandise was simply carried along the route and then another set was carried back to the beginning. That same system of cross-docking used in ancient trade is alive today. Companies like Wal-Mart have used this method of shipping to become very successful multinational corporations. Cross-docking saves time and money by cutting the amount of inventory required. The Internet has played a large part in the use of cross-docking. One firm in Turkey is still carrying on the cross-docking tradition that was started along the historical Silk Road. Ekol Lojistik, a 3PL company, helps other companies manage their logistics and helps to keep their costs down with innovative techniques like collaborating with other trucking companies that have less than full loads and getting their customers' goods shipped at a reduced price. Ekol also specializes in creating pallets of their customers' goods to increase shipping efficiency. The greatest advantage Ekol has is their information system that allows them to assist moving their customers' goods at the lowest possible cost.

In Chapter 9, Oliver Schneider, Robert Alard, and Josef Oehmen address the issue of changing buyer–supplier relationships between European and Chinese firms at either end of the Silk Road. As China grows faster than its Western counterparts, suppliers who were dominated by their Western buyers have slowly shed their shackles and begun to bargain as equals. Some partnerships have flourished even with these changes whereas others have stumbled. Through several case studies, the authors trace these developments. They also summarize the measures necessary to foster beneficial long-term relationship, including the important concept of Guanxi or special relationship. The authors conclude with an agenda for further research in this important area of collaboration along the Silk Road. It is interesting to observe that in the past, demand for Eastern goods was higher than demand for Western goods along the Silk Road due to specialized manufacturing skills (artisans) and the exotic features of goods. In a similar vein, today, many goods are manufactured in the eastern portions of the Silk Road and transported toward the West for consumption for somewhat different reasons than the past.

In Chapter 10, Qi Feng and Lauren Xiaoyuan Lu describe the progression of firms in the Asia-Pacific region from contract manufacturers to original design manufacturers (ODM) over the last three decades. Interestingly, as Feng and Lu point out, most of the ODMs are located along the eastern portion of the

Silk Road. They describe the economic forces that led to this gradual transition and how firms have adapted on both sides to the changing competitive landscape. They provide several success stories, including that of Dell, Google, and T-Mobile. Several academic studies are also summarized. These previous studies mainly focus on outsourcing and do not address the problems and prospects of design outsourcing specifically. The authors list a number of open questions related to design outsourcing, including those of product differentiation, competition from new entrants, quality, working conditions in the ODMs, environmental issues, etc.

In Chapter 11, Arif Iqbal Rana and Mohammed Kamran Mumtaz provide a detailed study of the supply chain for milk collection in Pakistan, a country at the crossroads of the Silk Road. They describe the current supply chains developed by Nestle that has adapted to social customs such as not selling milk during specific dates based on religious customs. Milk collection from farms, consolidated processing and packing, and distribution to retailers and then to customers provides several challenges. The role of payment schemes for milk, incentives for loyal farmers who supply steadily, and milk districts who do the same, all provide a glimpse of the management challenges that can be overcome to create successful supply chains. It is interesting to observe that not selling milk during specific dates based on Sufi tradition is a centuries-old tradition in Pakistan. Even though it sounds extraordinary in today's world, along the historical Silk Road, such traditions were accommodated in multicultural, multireligious, and multilingual environments.

In Chapter 12, Paul Lacourbe provides insights into the rail system in Hungary and its potential role in the new Silk Road linking East and West through train routes. Understanding Hungary's present rail system requires a historical perspective of the territorial changes of modern Hungary as well as political choices made by successive Hungarian governments. Being a landlocked country, Hungary is at the crossroads between trade flows from the East to the West as well as from the South to the North. The trade flows to major centers around Hungary provide a glimpse of the potential that can be unleashed if a new logistics rail corridor opens up. Will the EU and China work to enable Hungary to play such a role in the future Silk Road? Only time will tell.

In Chapter 13, Orla Stapleton, Lead Stadtler, and Luk Van Wassenhove focus on humanitarian supply chains along the Silk Road—the region has seen a growing number of catastrophes and demand for humanitarian assistance. They focus on private sector partnerships in such contexts. Such roles vary from a supplier role to a donor role. They focus on three examples of such partnerships involving Agility, Aramex, and TNT. They then provide a summary of lessons learned from these cases that can serve as a guide to developing effective humanitarian supply chains along the Silk Road. In historical Silk Road supply chains, public–private partnerships were used to manage supply chain risks. Governments such as the Seljuk Sultanate of Anatolia created a state insurance policy to mitigate the security risks of land and maritime traders due to bandit, pirate, and neighboring state attacks. Moreover, secure, safe, and smooth flow of merchandise, labor, and supplies was achieved through various instruments along the Silk Road. Garrisons and watchtowers beyond the Great Wall, postal stations, and caravanserais were built by Chinese, Mongols, and Turks, respectively. These structures also provided humanitarian aid when necessary to the traders. In sum, today's humanitarian supply chains along the Silk Road could learn a lot from historical practices.

In Chapter 14, Barış Tan presents a risk-based planning method for an agribusiness firm that supplies premium fruits and vegetables by using contract farming in Turkey. In the proposed approach, area and timing of seeding decisions are optimally determined while maximizing the expected profit. Long lead times and uncertainties in demand, supply, yield, harvest, and maturity make these problems challenging. Tan also provides a specific case study of Alara Agri Business, a firm located at Bursa (one of the major silk-producing cities of the historical Silk Road), Turkey, that produces and exports fresh cherries and figs to 22 countries across 5 continents. Turkey is the global leader in cherry production. It is shown that for cherry supply chains, the proposed method could increase the expected profits by 17%. Tan concludes that along the Silk Road countries, the agricultural contract farming industry can benefit from such an approach where weather and climate change risks abound. In historical Silk Road supply chains, agriculture of specific items was critical, such as rice, sugarcane, and mulberry. Demand, supply, and price risks were prevalent. Similar to today's contract farming concept, firms in the past acted as brokers between the peasants and European consumers in order to match supply to demand. Besides, peasants were flexible enough to convert the acreage from one item to the other depending on the market demand. Merchants also constructed networks of suppliers for reliable procurement, especially in India. Thus, it is interesting to note that many risk management practices in today's supply chains did exist centuries ago.

In Chapter 15, Muhittin H. Demir, Burcu Adıvar, and Çağrı Haksöz present the case study of Renkler Makina (RM), a Turkish firm in the machinery industry, located at İzmir (an important port city in the historical Silk Road), Turkey, and its procurement risk management practice. Being simultaneously a supplier to global firms such as Schneider Electric and Areva and procuring from small- and medium-sized suppliers creates a great supply chain management challenge for RM along the Silk Road. The authors discuss the current supplier evaluation, selection, and management practices, as well as the risk management philosophy of the firm. RM's supplier management practice consists of constructing an objective risk assessment tool and managing a fruitful relationship with suppliers for future sustainability. The practice of evaluating suppliers and working with an effective portfolio of suppliers was observed along the historical Silk Road supply chains, especially for silk and sugar production. However, due to long lead times and transportation constraints, unlike today's supply chains, supplier portfolios did not span many continents.

In Chapter 16, Hoda Davarzani and Andreas Norrman present the case study of an automobile firm and its supply risk management strategy in Iran. The region covered by Iran was served by the historical Silk Road supply chains. The authors first classify the supply chain risks of the firm into five groups: customer-, competitor-, supplier-, environment-, and organization-driven risks. Second, they focus on the alternating/hybrid sourcing strategy that would potentially mitigate supply disruptions. Specific circumstances are identified where such a strategy would be most beneficial for firms along the Silk Road.

We hope that the reader benefits from learning about centuries-old supply chain traditions and practices along the Silk Road and visualizes the connections to the present. Global trade along the Silk Road created the first supply chains of the world. And today, the great grandchildren of the Silk Road countries are revitalizing the international trade of goods and services along the Silk Road at an unprecedented scale and speed. This book celebrates past and present trends in supply chain management along the ancient Silk Road.

Çağrı Haksöz
İstanbul, Turkey

Sridhar Seshadri
Austin, Texas

Ananth V. Iyer
West Lafayette, Indiana

Acknowledgments

A book cannot be written without the support of many individuals. We would like to acknowledge those who supported us throughout the publication process. From the initial conception till the final stages, Nakiye Boyacıgiller, the dean of Sabancı School of Management, Sabancı University, played an important role in this project. We are grateful to her for her encouragement and support. Second, the academic environment created by Tosun Terzioğlu (former president of Sabancı University), as well as by the current president Nihat Berker, was instrumental in pursuing this project. We would like to thank Christopher Tang of UCLA-Anderson School of Management for writing a foreword despite his hectic schedule. We are also grateful to many of our colleagues, especially Michael Pinedo, Murat Kaya, Gürdal Ertek, Deniz Özdemir, Alfonso Pedraza Martinez, Gül Berna Özcan, Erhan Kutanoğlu, Metin Çakanyıldırım, Vinod Singhal, and the International Supply Chain Risk Management (ISCRIM) network that played critical roles at different stages of the book production process. We would like to thank Erkan Kusku, who designed our book cover as well as the Silk Road map in his unique style. We would also like to acknowledge the research assistance of Damla Durak Uşar, Ferhat Zor, Birkan İçaçan, Ersin Hasan Yörük, and Austin New. Our editors Lara Zoble and Marsha Pronin provided the necessary and timely support whenever needed.

More importantly, we are grateful to all of the contributors to this book across the globe for providing their invaluable insights and experiences. Without them, this book would never have existed. And finally, we would like to thank our families, whose support and love kept our spirits high. This book is dedicated to you all.

Editors

Professor Çağrı Haksöz received his PhD and MPhil in operations management from New York University, Stern School of Business, and his BSc in industrial engineering from Bilkent University, Turkey. He is currently a professor of operations management at Sabancı School of Management, Sabancı University, İstanbul, Turkey. His recent research has focused on supply chain risk management, design and valuation of options in supply chain contracts, product recall management, pricing weather derivatives, and empirical studies in supply chain management. His published works have appeared in *Operations Research, MIT Sloan Management Review,* the *Journal of Operational Risk,* the *Journal of Operational Research Society,* and the *International Journal of Production Research* among others. Professor Haksöz has taught undergraduate, MBA, and EMBA programs in the United States, the United Kingdom, and Turkey, as well as innovative executive programs such as Myglobe—Managing Your Global Enterprises (codeveloped by INSEAD) and TURQUALITY®. He has also served as a consultant for global and national firms. He currently serves as an associate editor for *Transnational Marketing Journal.* Before joining Sabancı University, Professor Haksöz was a faculty member at Cass Business School, City University of London, United Kingdom. Between 2008 and 2011, he was an honorary visiting fellow at Cass Business School, City University of London, United Kingdom.

Professor Sridhar Seshadri received his PhD from the University of California, Berkeley, California, after graduating from the Indian Institute of Technology, Madras, India, and the Indian Institute of Management, Ahmadabad, India. He is currently a professor of information, risk, and operations management at the University of Texas (UT). He is the fellow of the Center for Excellence in Supply Chain Management as well as the CBA Faculty Fellow at UT Austin. He has also been a faculty member at New York University and the Administrative Staff College of India. During his teaching career, he was awarded the Stern School of Business Teaching Excellence Award in 1998 and recognized as the Stern School of Business Undergraduate Teacher of the Year in 1997. His current research projects focus on equilibrium asset pricing, pricing and revenue optimization, and risk management in supply chains. In addition to teaching, he also serves on the board of directors for Nomi Networks and is on the advisory board of RSG Media, United States. His professional experience includes serving as the associate editor of *Naval Research Logistics*; area editor of *Operations Research Letters*; and senior editor (supply chain and stochastics) of *Production and Operations Management Journal*. He recently coauthored the book *Toyota's Supply Chain Management: A Strategic Approach to Toyota's Renowned System*.

Professor Ananth V. Iyer received his PhD from the school of Industrial and Systems Engineering at Georgia Tech, his MS in industrial engineering and operation research from Syracuse University, and his BTech in mechanical engineering from IIT Bombay. He is the Susan Bulkeley Butler Chair in Operations Management and the director of DCMME (Dauch Center for the Management of Manufacturing Enterprises) and GSCMI (the Global Supply Chain Management Initiative) at the Krannert School of Management at Purdue University. Previously, he was Purdue University Faculty Scholar from 1999 to 2004. His teaching and research interests include operations and supply chain management. Professor Iyer's research currently focuses on analysis of supply chains, including the modeling of spare parts supply chains, auto industry supply chains, the impact of promotions on logistics systems in the grocery industry, analysis of the impact of competitors on operational management models, and the role of supply contracts. His other topics of study include inventory management in the fashion industry, effect of supplier contracts, and use of empirical data sets in operations management model building. His published works have appeared in *Operations Research, Management Science, M&SOM Journal, EJOR, OR Letters, Networks*, etc. He was the FMC Scholar in 1990–1991. He has served as the department editor of *Management Science*; is an associate editor of *Operations Research*; is on the editorial boards of *Operations Research Letters, IIE Transactions*, the *ECR Journal*, and *Manufacturing and Service Operations Management*; and is a member of INFORMS (Institute for Operations Research and Management Sciences). He was formerly a senior editor at the *POMS Journal*. He was president-elect of the MSOM Society of INFORMS in 2001–2002 and served as president for the year 2002–2003. Prior to joining the Krannert faculty in 1996, Professor Iyer taught at the University of Chicago. He has been affiliated with the Production and Distribution Research Center at Georgia Tech, and serves as a consultant to Daymon Associates, Sara Lee, Turner Broadcasting, Dade Behring, Case New Holland, and others. He served his Chicago community as a pro bono consultant to the Chicago School System and the Chicago Streets and Sanitation Department.

Contributors

Burcu Adıvar
Department of Logistics Management
İzmir University of Economics
İzmir, Turkey

Robert Alard
BWI Center for Industrial
    Management
Swiss Federal Institute of Technology
    Zurich
Zurich, Switzerland

Hoda Davarzani
Faculty of Engineering
Department of Industrial Engineering
Tarbiat Modares University
Tehran, Iran

Muhittin H. Demir
Department of Logistics Management
İzmir University of Economics
İzmir, Turkey

Gürdal Ertek
Manufacturing Systems Engineering
Sabancı University
İstanbul, Turkey

Qi Feng
McCombs School of Business
University of Texas
Austin, Texas

Debabrata Ghosh
Indian Institute of Management
Bangalore, India

Çağrı Haksöz
Sabancı School of Management
Sabancı University
İstanbul, Turkey

Ananth V. Iyer
Krannert School of Management
Purdue University
West Lafayette, Indiana

Murat Kaya
Manufacturing Systems Engineering
Sabancı University
İstanbul, Turkey

Ercan Korkut
Nokia Siemens Networks
İstanbul, Turkey

Paul Lacourbe
CEU Business School
Central European University
Budapest, Hungary

Andrea Lenterna
Maxi Editor s.r.l.,
Perugia, Italy

Lauren Xiaoyuan Lu
Kenan-Flagler Business School
University of North Carolina
Chapel Hill, North Carolina

Ehud Menipaz
Stern School of Business
New York University
New York, New York
and

The Ira Center for Business
Ben Gurion University
Beer Sheva, Israel

Mohammad Kamran Mumtaz
Suleman Dawood School of Business
Lahore University of Management
    Sciences
Lahore, Pakistan

Andreas Norrman
Department of Industrial
    Management and Logistics
Lund University
Lund, Sweden

Josef Oehmen
Engineering Systems Division
Massachusetts Institute of
    Technology
Cambridge, Massachusetts

Arif Iqbal Rana
Suleman Dawood School of Business
Lahore University of Management
    Sciences
Lahore, Pakistan

Oliver Schneider
BWI Center for Industrial
    Management
Swiss Federal Institute of Technology
    Zurich
Zurich, Switzerland

Janat Shah
Indian Institute of Management
Udaipur, India

Lea Stadtler
Haute Ecole Commerciale
University of Geneva
Geneva, Switzerland

Orla Stapleton
Humanitarian Research Group
INSEAD
Fontainebleau, France

Barış Tan
College of Administrative Sciences
    and Economics
Koç University
İstanbul, Turkey

Deniz Tura
Toll Cross Securities,
Toronto, Canada

Damla Durak Uşar
Sabancı School of Management
Sabancı University
İstanbul, Turkey

Luk N. Van Wassenhove
Humanitarian Research Group
INSEAD
Fontainebleau, France

SUPPLY CHAIN STRATEGY ON THE SILK ROAD

Chapter 1

Silk Road Supply Chains: A Historical Perspective

Çağrı Haksöz and Damla Durak Uşar

Contents

1.1 Introduction to Silk Road Supply Chains: Trade and Players

In this chapter, we discuss Silk Road supply chains from a historical perspective and provide interesting connections to modern day supply chain management. We think that the first supply chains of real significance naturally appeared and operated along the Silk Road. Although the supply chain concept was coined in

3

the twentieth century, Silk Road trade practically created first successful examples of supply chains in human history. Hence, understanding the international trade, participants, processes used, products/services traded, lead times, governments, opportunities, risks, and inter-border and inter-cultural issues will be beneficial to keep in mind while comparing today's global and virtual supply chains along the same road. The periods we considered in this historical perspective are mainly ancient, medieval, and postmedieval until the nineteenth century. The Silk Road shown in Figure 1.1 refers to the region mainly between current day's China and Turkey (traversing through India, Pakistan, Afghanistan, Central Asia, Iran, Middle East, and Azerbaijan). The map in Figure 1.1 provides all of the overland and maritime trade routes used over the centuries.

First, we provide a perspective on why the trade on the Silk Road happened. Second, we describe the key players in the trade that plied along the road. It is important to note that the elite at both ends of the Silk Road wanted to consume luxury goods to differentiate themselves from common people. Luxury goods were valuable because of their scarcity and/or requirement of specialized manufacturing skills and are associated with the exotic (Oka and Kusimba, 2008). In the premodern era, luxury goods were not only for consumption and pleasure of the privileged, but also a form of political currency. Luxury goods such as silk cloths, advanced weapons, and gemstones were an element in maintaining political power (Allsen, 1997). However, throughout time the Western demand for Eastern goods was much more significant than the Eastern demand for Western goods to facilitate trade along the Silk Road (Morineau, 1999a). In today's world, various luxury goods are available for anyone who is ready to pay premium prices.

The participants along the Silk Road changed over time. "Chinese, Yuezhi, Bactrians, Indians, and Sogdians" created the historical Silk Road in Central Asia in the first century BC. Different groups rose and fell through the ages. Groups that gained political and military power controlled the trade along the Silk Road. In the medieval times, when paddle trade—merchants traveling along their goods—was common along the Silk Road, merchants and agents were the main players. During the eleventh to fourteenth centuries, Silk Road trade on the Western front (from Persia to İstanbul), was mainly controlled by the Great Seljuk Empire. Trade diasporas were formed such as Sogdians in Tang China, Armenians in Safavid Iran, and Jews in West Europe. Despite being open since the ninth century, maritime routes became more significant in the sixteenth century due to technological improvements in the naval architecture. The long distance maritime trade was monopolized by the Iberians. After the seventeenth century, West European charter companies such as the English East India Company (EIC) and Dutch East India Company (VOC) operated as single connected distribution chains from producer to consumer (Curtin, 1998). Today, similar integrated distribution functions are provided by global supply chain service providers such as UPS, Fedex, and DHL.

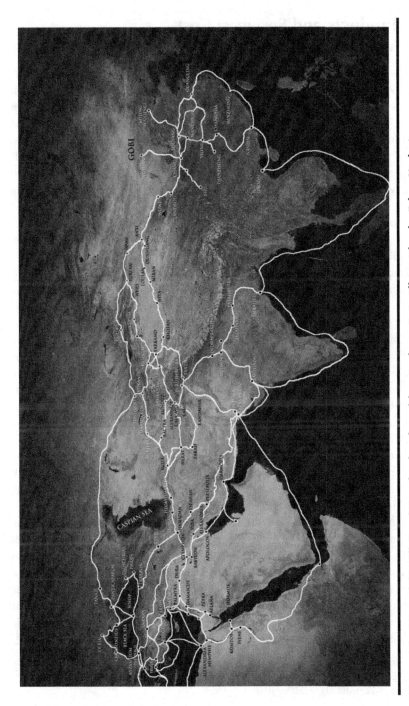

Figure 1.1   Map of the Silk Road (overland and maritime trade routes). (Illustration by Erkan Kusku.)

1.2 Products and Services: Cost, Variety, and Volume

Merchants sought significant profits so that it would be worth making the journey along the ancient Silk Road. Thus, goods traded overland had high value with respect to their bulk volume. The main good traded was silk from China. Other luxury goods included glass, red coral, furs, finely crafted ceramics, elegant bronze, jade, lacquer, iron, satin musk, rubies, diamonds, pearls, and rhubarbs. The goods traded along the Silk Road varied depending on who was controlling the routes at that particular time. In the fourth century, letters written by Sogdians were deciphered in Chinese Central Asia referring to commodities such as gold, wheat, pepper, musk, camphor, and flax cloth. The Sogdians managed a trade triangle between China, India, and Sogdiana,* which is located in modern day Uzbekistan and Tajikistan (Wood, 2004).

Trade along the Silk Road was at its zenith during the Tang dynasty due to the stability of the government. The most demanded foreign good on the far end of the Silk Road were horses during the Tang dynasty, since horses provided mobility and military advantage. The Chinese agricultural society was dependent on the Western regions for the supply of animals. The Uyghur nomadic society producing livestock demanded silk. The motivation of the Uyghurs for trade with the Chinese was not only the high demand for silk, but also the opportunity of horse trade in return (Liu, 2001). In some sources it is stated that the trade of silk for horses was a forced one, namely the Uyghurs benefitted from the economic advantage in return for their military assistance to Tang China (MacKerras, 2000).

During 820–830 AD, 500,000 pieces of silk per year were exported to the Uyghur Turks in exchange of horses. The price of 1 horse was 40 pieces of silk (MacKerras, 2000). The annual import of Uyghurs amounted 1 million bolts of silk in return of 100,000 horses (Liu, 2001). We can infer from these figures that 1 horse was worth 8–10 bolts, and one bolt contained approximately five pieces of silk. On the other hand, Romans demanded silk yarn and plain silk textiles instead of value added silk textiles, to be able to transform them into garments for priests, hangings for churches and coverings for saints. The price of one bolt of silk also varied based on location. It was around 200 copper coins in Central China, yet it cost 450–470 coins in the Western frontier of China, indicating even higher prices abroad (Liu, 1996). Hence, geographic locations created price arbitrage. The coinage alloy was 83% copper, 15% lead,

---

* http://en.wikipedia.org/wiki/Sogdian_language

and 2% tin. One thousand copper coins weighted 6 jin and 4 liang.* Thus 1000 copper coins weighed approximately 3781.25 g.

Access to different routes on land and ocean was critical along the Silk Road. In the eighth and ninth centuries, maritime routes (shown in Figure 1.1) gained importance compared to the dangerous overland routes. However, the overland trade routes were not completely abandoned, since caravans could access regions that are inaccessible to ships (Palat and Wallerstein, 1999). The developments in technology allowed ships to carry bulk commodities over great distances at lower costs. The opening of the sea based trade routes changed the types of traded commodities. The main export of China became porcelain along with spices, medicines, and timber (Lewis, 2009). It is interesting to note that, maritime cargo shipping, which has helped the global world become flatter today, was used centuries ago.

In the eleventh and twelfth centuries, due to the improvement of maritime technologies, it was possible to carry bulk cargo over long distances with less labor cost. Goods like pepper, spices, rice, sugar, wheat, salt, manufactured goods, and raw materials were traded (Curtin, 1998).

In the period of the Great Seljuk Empire (1037–1194) and the following Seljuk Sultanate of Anatolia (also known as Seljuk Sultanate of Rum) (1077–1307), east–west and north–south trade routes passing through Anatolia that connected Tabriz, Baghdad, Damascus, and Aleppo to the Western cities such as İstanbul and İzmir flourished. These routes constituted the Western (Persia, Asia Minor) sections of the Silk Road that travels via many branches in Anatolia. (Refer to the map in Figure 1.1.) At the Mediterranean, one branch began from Antalya, Alanya and traveled toward the Sultanate capital Konya, and then Kayseri, Sivas, and Erzurum. Another branch traveled from Erzurum-Sivas-Tokat-Amasya toward the Black Sea to Sinop. A third branch traveled from Konya to İstanbul via Afyonkarahisar, Kütahya, Eskişehir, and Bilecik. Yet a fourth branch coming from Tabriz, passed along Ağrı, Erzurum, and Bayburt and reached Trabzon port at the Black Sea. All these routes were controlled and secured by the Seljuk Empire. The Seljuk Empire originated from the *Qynyq* branch of the Oghuz Turks who controlled a large geographical area from Central Asia to Anatolia as well as Persian Gulf.

During the thirteenth century, caravanserais (rest houses) were built along the trade routes in Anatolia (Yavuz, 1997) mainly to serve the caravans of the traders. Caravans with their animals were able to stay in these caravanserais free up to 3 days. Food, accommodation, and health care services were provided.

---

* Jin and Liang are both traditional Chinese unit of mass and equal to 605 g and 1/16 jin, respectively. See for details http://en.wikipedia.org/wiki/Chinese_coin#Tang_issues and http://en.wikipedia.org/wiki/Catty, http://en.wikipedia.org/wiki/Tael

Even shoes were provided to the poor (Turan, 2009). In the Seljuk Empire period, these caravanserais were known as *han* or *derbend*. Taking into account the 1 day travel distance of the caravans, caravanserais were strategically positioned on the trade routes at 25–40 km away from each other. Sometimes the topography and the winter–summer daylight differences affected the location of these caravanserais (Eravşar, 2011). It is also known that the role of caravanserais were not only hosting caravans, but also serving as military stations, royal guesthouses for sovereigns, prisons, places for refuge, and religious meeting points (Yavuz, 1997). Besides building and maintaining the caravanserais, the Seljuk Empire also repaired old bridges and built new ones in order to enable a smooth flow of trade in and out of Anatolia (Eravşar, 2011). In sum, a smooth, secure, and safe flow of merchandize, labor, and supplies was enabled by building a network of caravanserais along the Silk Road.

During the Seljuk Sultanate of Anatolia, the Mongol Empire emerged and became a dominant military power in the thirteenth century. The Mongols formed a large empire that covered the Chinese land in the east to Persian land in the west. Thus, they mostly controlled the intercontinental trade. The stability and security provided by the single authority uniting such a wide land mass stimulated trade along the Silk Road (Curtin, 1998). The Mongolian demand for golden embroidered silk was high. The volume of silk fibers consumed was 425 and 655 ton in the years 1263 and 1368, respectively. It is interesting to note that the main supply of gold brocade was not through trade. The major amount of gold brocade entering the Mongol Empire was loot and collected taxes from invaded states. As Mongols had high regard for gold brocade, the invaded states paid their taxes in forms of gold brocade garments. At that time, the price for a garment of gold brocade was one gold ingot (Allsen, 1997).

In the commercial handbook written by Francesco Balducci Pegolloti in the fourteenth century, the expenses for overland trade are provided. The expenses of caravans transporting goods of 25,000 golden florins value would be 3,500 golden florins. The transportation costs including labor, animals, and supplies would be 1000–1500 florins assuming 60 men were employed and 40–60 animals were used to carry the goods. Total duty paid would be 1600 golden florins, 400 for round-trip duties and 1000 for customs, assuming the charge was 5% of the value of the goods on average. Payment demanded for permission to travel through the territories of various states would be 200–300 florins assuming five florins per pack animal (Rossabi, 1990).

The Mughals gained power in India in the sixteenth century and controlled the trade along the Silk Road. During the sixteenth to eighteenth centuries, the most important markets of Mughal India were the countries bordering the Indian Ocean: Safavid Iran and the Central Asian steppe (Dale, 1994). Mughals originated from Central Asia, thus demanded Central Asian fruits such as apples, pears, grapes, plums, and melons and birds such as falcons, hawks, and pigeons. Furthermore, the main commodities imported to Mughal India were horses. Around 100,000 horses per year were traded without using cash, but in exchange for cotton, indigo, and sugar (Foltz, 1998).

In the sixteenth century, the Iberian empire controlled the sea trade from Asia. The main good imported was pepper, while the main good exported was copper. Iberian merchants had a huge price advantage in pepper trade over their rivals that were able to provide copper. The value of copper was from 2.5 to 4 times its weight in pepper, thus the net profit of importing pepper was 89%–152%. Besides pepper, other spices such as ginger, cinnamon, cloves, nutmeg, and mace were imported as well. Due to the small volumes, the pepper trade was not significant despite higher unit profits (Phillips, 1990). Besides copper, the Iberian Empire also traded coral and silver with Indian Mughal Empire (Morineau, 1999b).

During the sixteenth to the eighteenth centuries, three alternative trade routes linked the Ottoman Empire to India (refer to Figure 1.1 for these routes). Two of them were maritime routes, which were the Persian Gulf route and the Red Sea route, while the overland route traversed Anatolia. The Ottoman Empire imported various kinds of Indian fabrics, raw cotton, spices, indigo, iron, and steel from India and exported horses, nutgalls—a herb containing tannin used to cure intestinal disorders, bleeding, and excessive discharge in Chinese medicine*—buffalo skins, dates, rice, henna, camels, and madder dye—a vegetable red dye for leather, wool, cotton, and silk. The Ottoman demand for Indian goods was larger than the Indian demand for Ottoman goods. Hence, a net cash flow from Ottoman Empire to India was seen. The consumption of the Ottoman Empire could not be clearly estimated since the Empire was a transit zone to the West and part of the goods entering the empire would have been reexported. Yet the consumption of the Indian textiles in İstanbul was 66.5% of the consumption of the Levant (Veinstein, 1999). Similarly, modern day Turkey acts as a transit hub for many products that travel east–west and north–south directions.[†]

On the other hand, as stated by Matthee (1999), the trade figures between Persia and India in the seventeenth century can be estimated from the annual volume of camels passing via Qandahar (a city in today's Afghanistan) and the volume of camels arriving in Isfahan, which varied throughout the years. 7,000–8,000 camels (circa. 1610) and 12,000–14,000 camels (circa. 1615) passed through Qandahar, while 20,000–25,000 camels (circa. 1639) and 6,000 camels (circa. 1644) arrived in Isfahan loaded mainly with Indian cotton along with indigo (an organic blue dye). The caravans traveling in the opposite direction carried silver of the West and letters for the English company. The goods that

---

* See for example, Dharmananda (2003) for more details.

† Also see Chapters 7 and 8 in this book for 3PL industry examples in Turkey.

traveled via the ocean route were more diverse. Transportation costs including custom duties did not differ much based on the transportation mode. Along the overland route, transportation costs included 3% for customs and 10% along the ocean route. The annual import of Persia from India is summarized in Table 1.1.

In the seventeenth century, silk production in Iran was approximately 8000 bales per year depending on the warfare in the country and factors affecting the harvest such as climate, aridity, and disease. The weight of one bale ranged from 70 to 90 kg silk. Although precise information about the proportion of silk exported is not available, we can infer from diverse data available that one-half to three-quarters of the production was exported. Silk was purchased at 50 tumans and sold at 85 tumans, while transportation costs were 13–14 tumans

Table 1.1 Annual Import of Persia from India in 1634

	Quantities	Purchase Price in 1000 Rs.
Overland		
Cotton textiles	211,000 pieces	541
By ocean route		
Cotton textiles	383,000 pieces	841
Indigo	200,000 ponds[a]	244
Sugar	966,000 ponds	94
Gumlac	227,700 ponds	145
Pepper	138,000 ponds	14
Ginger	276,000 ponds	11
Tin	34,500 ponds	9
Cardamom	34,500 ponds	9
Total	18,767,000 ponds 594,000 pieces	1908

Source: Steensgaard, N., *Companies and Trade: Europe and Asia in the Early Modern Era*, Cambridge University Press, Cambridge, U.K., 1999.

[a] Steensgaard (1999) states that a standard camel can carry 450 lb or 412 ponds (avoirdupois pound, which was used as a unit of mass between the fourteenth and eighteenth centuries).

(Matthee, 1999). The value of 30,000 tumans was equivalent to 1,730 kg of pure gold (Hinz, 1969). In the beginning of the eighteenth century, 130 ton of silk produced in Ghilan, a province lying along the Caspian Sea, were annually exported. About 20% of the export went to the Russian market and the rest moved to the West European market. In the mid-century, the Ghilan production dropped to 110 ton. Twenty percent of the production was consumed in the domestic market, while the rest was exported (Curtin, 1998).

National monopolies such as EIC and VOC emerged in the seventeenth century. Pepper, spices, textiles, and silk comprised 75% of the imported goods. With the changing preferences of the consumers, coffee and tea became the most demanded Asian imports. The different commodities imported annually by the EIC and VOC are summarized in the Tables 1.2 through 1.4.

The detailed shipping costs of VOC in the eighteenth century are also available. The freight prices were at their minimum between the years 1716 and 1740. The prices were around 24–30 pounds per ton in this period, while above 35 pounds per ton throughout the eighteenth century (Brujin, 1990). Details are provided in Table 1.5.

In the eighteenth century, the northern trade route became an alternative to the route linking China, Uyghur Region, Persia, and the Middle East. The northern route traversed Siberia and northern Central Asia, which was a Russian territory. The custom duties and expenses for defense were low due to stability provided by a single country. Russian merchants bought cotton, tea, tobacco, and rhubarb from the Chinese and sold in return furs, leather, woolen, weapons,

Table 1.2 Annual Average Imports from Asia

VOC			
	1664–1670	Invoice values in 1000 pesos	980.8
		Sales values in 1000 pesos	3515.1
	1670–1700	Invoice values in 1000 pesos	1520.63
		Sales values in 1000 pesos	4063.2
EIC	1644–1670	Invoice values in 1000 pesos	437.2
		Sales values in 1000 pesos	1184.9
	1670–1700	Invoice values in 1000 pesos	4063.2
		Sales values in 1000 pesos	2973.1

Source: Steensgaard, N., *The Rise of Merchant Empires: Long Distance Trade in the Early Modern World, 1350–1750,* Cambridge University Press, Cambridge, U.K., 1990.

Table 1.3 Distributions of Imports by Commodities

	VOC		EIC	
	Invoice Values in %	Sales Values in %	Invoice Values in %	Sales Values in %
	1664–1670			
Spices	12.05	28.43		
Pepper	30.53	28.99	20.01	19.46
Sugar	4.24	2.02		
Tea and coffee		0.03	0.65	0.63
Textiles and silk	36.46	23.77	63.07	57.84
	1670–1700			
Spices	11.70	24.78		
Pepper	11.23	13.31	6.14	10.48
Sugar	0.24	0.20		
Tea and coffee	4.24	4.10	2.79	3.22
Textiles and silk	54.73	43.45	78.09	74.89

Source: Steensgaard, N., *The Rise of Merchant Empires: Long Distance Trade in the Early Modern World, 1350–1750,* Cambridge University Press, Cambridge, U.K., 1990.

and utensils made from iron. The total volume of the trade was 837,066 rubles in 1755 and increased to 8,383,846 rubles in 1800 (Rossabi, 1990).

The French Indian Company also engaged in shipping silver, metals such as iron, copper, and lead, as well as, alcohol to Asia. The value of the silver shipped was 75%–85% of the total cargo. The value of the silver fluctuated between 14,652,851 and 64,336,277 Livre tournois (Lt)* during the years 1725–1769. The amount of flat iron shipped annually was 600,000 lb, while the amount of silver varied from 13,615 to 51,670 ton. It is surprising that the amount of silver shipped to India was greater than the amount of iron. European consumers

---

* The Livre tournois abbreviated as Lt is one of the currencies used in France in the middle ages and is worth about $4 per livre in today's money. Available at http://en.wikipedia.org/wiki/Livre_tournois

Table 1.4 Annual Quantities and Sales of Asian Imports in 1752–1754

	VOC		EIC	
	Quantities in Pieces and 1000 kg	Sales Values in 1000 Pesos	Quantities in Pieces and 1000 kg	Sales Values in 1000 Pesos
Spices	487	1852	0	0
Pepper	1679	677	1,061	404
Tea and coffee	2608	1791	2,046	1971
Textiles and silk	Not available	2157	604,156	4853

Source: Steensgaard, N., *The Rise of Merchant Empires: Long Distance Trade in the Early Modern World, 1350–1750*, Cambridge University Press, Cambridge, U.K., 1990.

demanded Indian goods while Indian consumers were not equally interested in European goods. Hence, Europe tried to pay their import with the revenues from the trade of the goods they could offer to the other end of the Silk Road. However, they had to cover the major part of their expenditures with silver. Thus the asymmetry in demand led to increasing figures of silver. The alcohol shipped to Asia mainly catered to Europeans living abroad (Haudrère, 1999).

Europeans imported porcelain, lacquered ware, silk, cotton cloth, refined and raw sugar, and tea during the eighteenth and the nineteenth centuries. The Europeans had to pay with silver due to lack of Chinese demand for European commodities. However, the EIC managed to trade Chinese tea with raw cotton imported from India. The silk imported to Europe amounted 25,000 piculs and increased to 1.1 million piculs* in 1828 (Marks, 1998).

1.3 Supply Chain Processes: From Procurement to Sales

In the past, silk production was very labor intensive and required important manual skills. The production process consisted of *sericulture, reeling,* and *throwing.* Sericulture is rearing of silkworms for the production of raw silk.

---

* Picul was a Javanese unit of weight approximating 61 kg. Spanish, Portuguese, British, and Dutch colonial maritime trade companies employed picul in their transactions, since the unit was widely understood and used by Indonesian, Javanese, Chinese, Indians, and Arabs. Available at http://en.wikipedia.org/wiki/Picul

Table 1.5 Shipping Costs of VOC in the Eighteenth Century

	Materials (f)	Victualing (1000f)	Wages (1000f)	Total (1000f)	Price per Ton (1000f)	Tonnage
1700–1710	15,444	9,142	29,608	54,194	291	186,364
1710–1720	17,629	10,238	31,830	59,697	262	228,066
1720–1730	22,438	11,816	37,846	72,100	249	289,233
1730–1740	25,325	12,022	39,196	77,263	276	280,035
1740–1750	21,349	14,006	36,516	71,871	284	252,715
1750–1760	22,841	11,156	42,733	76,730	275	278,845
1760–1770	22,525	11,223	44,626	78,374	269	291,605
1770–1780	22,934	12,479	44,032	79,445	274	290,340
1780–1790	34,353	1,263	39,986	86,975	357	243,424

Source: Bruijn, J.R., *The Rise of Merchant Empires: Long Distance Trade in the Early Modern World, 1350–1750*, Cambridge University Press, Cambridge, U.K., 1990.

The silkworms are fed mulberry leaves. The silkworms are ready to spin the silk cocoon about 25 days old. The silkworm encloses itself in the cocoon in about 2–3 days. Afterwards, the cocoons are immersed into hot water. The silkworm pupae dies and approximately 1000 yards of silk filament per cocoon are obtained.* The silk filaments obtained from the cocoons are winded on reels. Silk that has been reeled into skeins is cleaned, twisted together with threads, and wound onto bobbins.† Sericulture depended on climate, because mulberry growing and silkworm raising requires a certain temperate and subtropical climate. Thus sericulture process was not widespread. Only after the nineteenth century, it became economically feasible due to technological progress to maintain the right conditions for sericulture artificially. This enabled production independent of the climate. The reeling process took place in sericulture areas since the input of the process were cocoons and transporting the cocoons was costly. Then, the labor cost became the most important concern in locating reeling factories. Since the reeling process was the highest value adding process, its location could be decoupled from sericulture and the throwing process. Reeling was generally carried out in commercial centers to reduce transaction costs. With the development of railway transportation, reeling process was located in cities where the supply of manpower was high and wages were reasonably low (Federico, 1997). Procurement of silk in seventeenth century Iran was a laborious operation, since silk was collected bundle by bundle from each cultivator. The purchase of 100 bales of silk required the work of at least 20 employees (Matthee, 1999).

In history, few people traveled the full length of the Silk Road. The goods were transported by a series of routes and agents. This structure reminds one of today's multiagent, multimodal global supply chains. The routes used by the merchants were selected according to the political stability and the topography. The Silk Road lies on the most challenging geography of the world in terms of vast deserts and high and rugged mountain ranges. The extreme continental climate, great distances between human settlements, lack of governmental administrations, and banditry were the main problems travelers had to face. Travelers joined caravans to benefit from safety in numbers and experience of the caravaneers gathered from their previous trips. Travelers transported their goods and personal belongings mainly on horses, donkeys, and mules. Yet they had to adjust the type of animal according to the stage of the journey. If the caravan was passing the desert, camels were used. In the colder and higher elevations, dromedaries and yaks were used. The amount of goods carried and the pace set by different types of animals varied (Foltz, 1999). For example, the Bactrian camels traveled at about 2.5 miles an hour. The cost of a camel in its

---

* http://en.wikipedia.org/wiki/Sericulture
† http://en.wikipedia.org/wiki/Silk_throwing

prime was 14 bolts of silk. The hirers were responsible for death/injury risks of any animal during the hiring period. It is noted by Wood (2004) that threat of sandstorms was great on some routes. Old camels were experienced enough to warn the rest of the caravan members in such cases. Goods were carried in packs that weighed up to 200 lb each (Whitfield, 1999). Furthermore, the Arabian camel could carry two bales of silk, which is approximately 180 kg. A mule could carry 150–180 kg while an ass could be loaded with 100 kg (Matthee, 1999). In sum, transportation strategy along the Silk Road was flexible in terms of route and mode selection. Topography, riskiness of the routes, and the climate were considered in detail.

The transportation costs consisted of freight prices, road tolls, and custom fees. These costs fluctuated according to the season, state of the roads, expected travel time, and availability of animals and fodder (Matthee, 1999). Overland trade as well as maritime trade was cyclic and seasonal, due to natural circumstances. Weather conditions in the passes of Central Asia, monsoon system in the Indian Ocean, and the seasonal winds and currents in the Mediterranean played a great role in the seasonality of trade (Jacoby, 1997).

In the sixteenth century, joint ventures of merchants and the Portuguese Crown took place. Their terms dictated equal share of the cargo space aboard and expenses of the voyage. Individual merchants also joined the voyage by paying freight charges for cargo space occupied by their goods. Moreover, merchants participated in the venture by handing out money without undertaking the actual journey (Prakash, 1999). This centuries old practice has been honed into cargo capacity allocation/apportioning strategy of today's global cargo carriers.

In the second half of the sixteenth century, the Portuguese Crown implemented a concession system instead of taking part in transactions. Thus, the Portuguese Crown extracted profits without taking any risks connected with the market. The concession system gave exclusive right not only to trade on the route, but also collect freight charges and custom duties to private entrepreneurs (Prakash, 1999). Moreover, the position of captain-major was given to those that had civil and criminal jurisdiction over all people on the trading fleet (Subrahmanyam, 1997). The concession was given as a reward to nobles or soldiers for their services. A concession was sold to the highest bidder (Prakash, 1999). There were 34 annual concession voyages in operation across the Bay of Bengal covering the South China Sea and the Indonesian archipelago (McPherson, 2004). The values of the concession voyages in 1580 varied from 8000 to 1000 cruzados (Subrahmanyam, 1997). The silver cruzado was worth 400–420 réis (plural of Portuguese currency real) during the sixteenth century.*

---

* http://en.wikipedia.org/wiki/Portuguese_real

The Dutch practiced the concession system as well. Their motivation however was different than the Portuguese. It was used as an instrument to regulate the competition (Gaastra, 1999).

In the seventeenth century, the transportation and distribution along the Silk Road drastically changed. In the medieval times, commodities were traded via multiple agents in city states and maritime republics. In the early modern era, trade did not take part in stages unlike the medieval times. In contrast, consumers and producers were directly linked by single agents. Commission agents and salaried employees of chartered companies operated along the Silk Road without any involvement of middleman, which increased cost efficiency (Matthee, 1999). Merchants did not need to travel with their goods any more. The goods were sent via different jurisdictions by employing a variety of transfer modes (Curtin, 1998).

It is interesting to note that supply chain contracts were designed and used 14 centuries ago. The documents excavated from Turfan, located in modern day Xinjiang Uyghur Autonomous Region, which was once of the largest trade centers on the Silk Road around 640, show that the locals and Silk Road traders had used contracts for goods purchased and money borrowed. Prices were clearly documented and the contracts included 6%–10% penalty charge/interest for each month of delay in shipment/payment. In addition to long-distance exchange contracts, travel passes were issued, since Tang law stipulated that a caravan owner had to provide documentary proof that he owned the slaves and animals traveling with him. Furthermore, labor contracts, such as contracts to hire someone to transport goods to a given destination were also used (Hansen, 2007).

In the battle of Talas between Uyghurs-Tibetan allied army against Chinese in 751 AD, Chinese paper makers were captured. This led to paper manufacturing in the city of Samarkand where raw materials (such as hemp, flax, water) existed abundantly. Later, via Samarkand, paper manufacturing moved westward along the Silk Road.

In the early days, a barter economy existed along the Silk Road. Around 580 AD, silver coins were introduced to the Silk Road economy. Silk Road traders borrowed silver coins at the interest rate of 10% per month from the moneylenders. In the eighth century, silver coins were replaced by bronze coins (Hansen, 2007). Later, in the eleventh century both Muslim and Christian merchants employed banking instruments for making payments remotely, from which we can draw a parallel with the modern bill of exchange (Curtin, 1998). In the seventeenth century Mughal India, money was transferred through money changers, merchants, and government officials. The service charge was about 1%–2%. The interest rates charged on bills of exchange were approximately 0.75% (Curtin, 1998). In the eighteenth century, Indian money changers demanded a much higher, that is, 12%, commission (Haudrère, 1999).

1.4 Lead Times along the Silk Road Supply Chains

During the Tang dynasty, one bolt of silk about 30 ft took 1 day's work by one person (Whitfield, 1999). In the nineteenth century, in Central China reeling plants employed 600 workers on an average and produced 20 ton of raw silk annually (Federico, 1997).

In the fifteenth century, traversing from Herat in Afghanistan to Peking in China took about a year. The journey from Peking to Samarkand also took about a year. The travel time from İstanbul to Central Asia was 9–10 months. The travel time of full length of the Silk Road was roughly 270 days, when delays due to weather conditions or other inconveniences and rest periods were excluded (Rossabi, 1990). In the seventeenth century, it took caravans 5–6 months to traverse the route between Isfahan and Agra passing through Lahore, Qandahar, Farah, and Yazd (Steensgaard, 1999).

On the Western front of the Silk Road, it took camel caravans 45–75 days to traverse the overland route from Mediterranean to the Persian Gulf. There were two alternative routes from Ghilan to West Europe in the eighteenth century, one leading through Astrakhan, Moskow, and Saint Petersburg taking 95 days and the other one destined to İzmir and Turkey taking 70 and 60 days, respectively (Curtin, 1998). The duration of caravan travel from Qazvin to İstanbul/İzmir was 110 days, from Qazvin to Aleppo 110 days, from Qazvin to Archangel 105–135 days, and from Isfahan to Aleppo 70 days (Matthee, 1999). In sum, lead times for transportation and distribution of goods were long and variable; the journeys had to overcome many perils.

1.5 Careers on the Silk Road Supply Chains

Some merchants served as agents while other merchants practiced paddling trade. Namely, they traveled with their goods along the Silk Road. A second group of merchants settled down abroad and learned the language and trade customs of the hosts and acted as cross-cultural brokers between the merchants of the host society and the merchants of their homeland (Curtin, 1998). Today, global supply chain service providers do operate as brokers and agents in many continents. Li and Fung is one successful example that orchestrates a multitude of global retailers and suppliers for fashion, furnishings, and household products.* Types and varieties of careers seen along the Silk Road can be listed as follows:

---

* See for example, Fung et al. (2007) for a detailed account of Li and Fung and its global supply chain operations.

- Producers of goods
- Skilled craftsmen
- Bankers
- Moneylenders and moneychangers
- Commission agents
- Carriers such as caravaneers, carters, muleteers ship owners, and ship captains and crew
- Interpreters and translators
- Caravan guides and guards
- Law and contract enforcement officers for business transactions
- Musicians, dancers, and courtesans

Hansen (2007) illustrates a case between a Chinese merchant and the brother of his deceased partner to address law and contract enforcement for business transactions. The brother brought a complaint before the authorities in Turfan that the Chinese merchant did not pay back the 275 bolts of silk he had borrowed from his deceased partner. The Chinese merchant denied borrowing anything from the Sogdian partner. Two Sogdian merchants witnessed the original loan of the 275 bolts of silk and gave a testimony against the Chinese merchant. Although the copy of the contract belonging to the deceased Sogdian partner had disappeared, the case was closed in favor of the Sogdian merchant. Because according to Tang law the testimony of witness had the same legal standing as a copy of the contract (Hansen, 2007).

Trade along the Silk Road also supported the economy of intermediate industries and locations that provided services to travelers, such as caravanserais (Weisbrod, 2008). Caravans needed guides over the most hostile sections of the journey who knew the land and the climate. Interpreters in the medieval times were often illiterate and translated oral communications simultaneously while translators translated written texts. They were more likely bilingual or multilingual on the native level by birth, who were familiar not only with the language, but also with the customs of both parties. The Byzantine Empire had contact with Turkish civilizations, Arabs, and Persians, thus bilingual or multilingual individuals who interpreted from Greek to Turkish, Arabic, or Persian and vice versa were required. With the decline of the Byzantine Empire, the commerce with the West increased, thus interpreters, who knew Latin, were more in demand.

The Tang China had political and commercial contacts with the people of Inner Asia, thus required interpreters who knew both languages. With the rise of the Mongol Empire in the thirteenth century, interpreters who knew the Mongol tongue were needed. The Mongols formed a large empire that was located from Chinese land to Persia. This area covered the territories of Uyghur

Turks, so Uyghurs became subjects of Mongols. The Uyghurs had commercial and cultural relationships with the Chinese before the Mongol Empire emerged. Having knowledge about the Chinese culture, customs, and language as well as being a Turkic speaking tribe under Mongol rule, Uyghurs were perfect to serve as middlemen in political and commercial affairs of the Mongols with the Chinese. On the other hand, Armenians served as interpreters at the Western border of the Mongolian empire (Sinor, 1982).

Entertainers such as musicians, dancers, and courtesans were in great demand along the Silk Road. For example, courtesans were paid 16,000 cash per evening (Whitfield, 1999). We should note that "cash" in this context does refer to "tangible currency," not the English word "cash" per se. It is an older word from Middle French "caisse." This word was derived from the Tamil kāsu, a South Indian monetary unit, a type of copper coin used in Tang China.*

1.6 Managing Risks along the Silk Road Supply Chains

First, the Silk Road supply chains had to manage various financial risks such as interest, tariff, and tax rate risks. In the seventh century, the Parthians and later Sasanians who served as middlemen tried to monopolize the silk trade. The Turkic states in Central Asia demanded high tariffs for passage (Frye, 1996). Moneylenders charged the same high rate of interest to Silk Road traders, as well as, to local cultivators (Hansen, 2007).

Second, there was a great deal of supply chain security risks along the Silk Road. Different types of mitigation tools and methods were designed and operated at firm and across firm levels. Defense against banditry took place at private and institutional level. Caravans of goods needed their own guards against plundering by the bandits (i.e., for security risk), and this was an added cost for the merchants making the trip. The institutional level had three forms: The Chinese garrisons and watchtowers beyond the Great Wall, Mongolian postal stations, and caravanserais in the Middle East and Anatolia. These institutions provided safety, supplies, and lodgings for merchants. Besides, the Chinese soldiers informed about incidents using smoke and flag signals in real time (Rossabi, 1990). The Seljuk Sultanate of Anatolia created a state insurance policy† in order to manage the security risks of land and sea traders whose goods are damaged or stolen due to bandit, pirate, and neighboring state attacks (Turan, 2009). For

---

* http://en.wikipedia.org/wiki/Chinese_cash_(currency_unit)
† In a similar vein, also see Chapter 13 in this book for successful public–private partnerships for humanitarian logistics along the Silk Road.

insurance purposes, contracts were signed between caravaneers and merchants that guaranteed the quantity of the goods and also reduction in the transportation fee if any delays occurred. Similar transportation contracts are used by third part logistics firms in today's supply chains.* The caravaneers kept lists of goods carried with specifications such as variety, weight, and volume (Matthee, 1999). This practice is the origin of today's bill of lading in global supply chains.

Third, players along the Silk Road were also exposed to supply and procurement risks in different regions for various products. For instance, European merchants were exposed to more risks in procurement from India than procurement from China. While various goods were accessible in trade centers in China, trade centers in India were not well supplied. Merchants had to search for goods demanded inside India and had to make sure that the goods were manufactured and shipped to Europe according to specifications (Haudrère, 1999).

Fourth, demand and price risks were also prevalent. In the eighteenth and nineteenth centuries, European demand for refined and raw sugar, as well as, silk grew in unexpected proportions, resulting in conversion of acreage devoted to rice paddy to sugarcane and mulberry trees (Marks, 1998). In the nineteenth and early twentieth centuries, steam-reeling firms operated widely in China and the Mediterranean, especially the city of Bursa in Anatolia. The challenge for the firms was to purchase the cocoons at low prices and sell the silk at reasonably high prices in order to be profitable. Firms also acted like cultural brokers between local peasants and European consumers. Foreign-owned firms in China had more disadvantages than their Mediterranean counterparts in handling with the local peasants due to cultural differences (Federico, 1997).

1.7 Roles of Governments along the Silk Road

In the sixth century, Byzantium controlled the import of raw silk to protect the monopolistic position of the state since they had their own sericulture process. In the ninth century, the Byzantine state encouraged the import of raw silk by private purchasers to support the production of silk garments (Liu, 1996). In the twelfth and thirteenth centuries, the Seljuk Sultanate of Anatolia applied a low customs tax to foreign traders, especially of Latin origin, in order to encourage international trade (Turan, 2009).

In the fifteenth century, a few states in the Indian Ocean and South China Sea provided services like warehousing and offered low duties to promote maritime trade and profit from it in the long run. Other states tried to profit from their military and political power by restricting trade (Curtin, 1998).

---

* See Chapters 7 and 8 that present two company examples from the Turkish 3PL industry.

European companies imported Asian goods and exported European goods in return. Yet the demand for European goods and Asian goods were not balanced. Thus, European companies had to trade with silver and gold. Around 1550, Portuguese traded with silver, too. Since the Portuguese Crown was able to cover the deficit with revenues from the Americas, initially this practice, that is, export of silver, was not considered problematic. Yet, after 1580 awareness against trading with silver was created. There were two aspects of silver export. The companies trading in silver had competitive advantage against companies trading with merchandised goods. The second aspect was more important in the long run. The outflow of silver was emptying the treasuries of France, England and alike. In 1615, for example, the silver shipped to Asia was worth seven million écus, which was "almost the third of the money of France and two-thirds of Spain" (Morineau, 1999b). In 1615, one écu was worth about USD 12.8 in today's money (Spooner, 1972). Hence, 7 million écus would amount to approximately USD 84 million in today's monetary terms.

1.8 Managing Inter-Border and Inter-Cultural Differences

In the sixth and seventh centuries, Sogdian merchants were trusted by Central Asian states rather than their Chinese counterparts due to their lack of political ambitions. Thus, Sogdian became the *Lingua Franca* along the Silk Road (Frye, 1996). In the tenth century, Sogdians were absorbed by Uyghur Turks, and Sogdian, which is a middle Iranian language that was spoken in Sogdiana, was extinguished. However, Uyghur Turkish did not become the commercial language along the Silk Road. Persian and Cuman,* a Turkic language spoken by the nomadic Cuman Turkish tribe, who lived in the Eurasian steppe and originated from east of Yellow river, China became the lingua franca along the Silk Road (Sinor, 1997). After the spread of Islam in western Central Asia, Arabic and Persian were used for business transactions (Frye, 1996).

In the fifteenth century, maritime trade in the Indian Ocean and South China Sea took place among merchants with diverse cultural backgrounds in port cities linked to inland caravan centers. Despite the lack of cultural homogeneity, the merchants operated with the common culture of trade. New entrants were accepted to the trade system with reference to their cultural origins and earned support from fellow countrymen for lodging and cross-cultural brokerage (Curtin, 1998).

---

* http://en.wikipedia.org/wiki/Cuman

In the seventeenth century Mughal India, merchants had to trade within the general framework of Islamic contractual law. However, the information whether the law applied to non-Muslims is not clear. Persian was widely spread as administrative and cultural language in the region, thus both Indo-Muslim and non-Muslim merchants used Persian in their transactions (Dale, 1994).

1.9 Conclusion

In this chapter, we have presented interesting factual details on historical Silk Road supply chains, an ancient trading route that was complex but resilient. It is clear that at all levels (personal, firm, multifirm, government), many overlapping supply chains operated along the Silk Road that learned from each other in order to create value. These supply chains were instrumental in producing, distributing, and selling a diverse set of products to the vast geographic region encompassed by the Silk Road, from China to the Balkans and Europe for a very long period in human history.

Successfully moving goods from East to West and West to East along the Silk Road was a challenging task that required many institutions to be developed. Creating a strategy to manage the performance and the risks was similar to today's global supply chains. However, twenty-first century's ubiquity of information and 24/7 pace of international trade did not exist. Hence, demand supply mismatch was not uncommon. On the other hand, states and empires have emerged and capitalized on the Silk Road operations to gain competitive advantages. Having observed the intricate details of the Silk Road supply chains, both spatially and temporally, we can conclude that the seeds of today's modern global supply chain concepts and ideas were planted many centuries ago. As the countries along the ancient Silk Road emerge as globally competitive entities in the new century, we believe it is time to learn from the past. The passage of time has permitted the seeds sown in ancient times to germinate and blossom into modern twenty-first century supply chains.

Acknowledgments

We are grateful for important suggestions and comments of Sridhar Seshadri and Ananth V. Iyer who have contributed substantially in the development and the structure of this chapter. The first author also thanks for the fruitful discussions with Mehmet Haksöz, Osman Çendek, İsmail Cem Atalay, Murat Kaya, Metin Çakanyıldırım, and Gürdal Ertek. We also appreciate the timely contribution of Erkan Kusku who prepared the Silk Road map.

References

Allsen, T. 1997. *Commodity and Exchange in the Mongol Empire: A Cultural History of Islamic Textiles*. Cambridge University Press, Cambridge, U.K.

Brujin, J.R. 1990. Productivity, profitability and costs of private and corporate Dutch ship owning in the 17th and 18th centuries. In Ed. Tracy, J.D. *The Rise of Merchant Empires: Long Distance Trade in the Early Modern World, 1350–1750*. Cambridge University Press, Cambridge, U.K.

Curtin, P.D. 1998. *Cross Cultural Trade in World History*. Cambridge University Press, Cambridge, U.K.

Dale, S.F. 1994. *Indian Merchants and Eurasian Trade, 1600–1750*. Cambridge University Press, Cambridge, U.K.

Dharmananda, S. 2003. *Gallnuts and the Uses of Tannins in Chinese Medicine*, Available at http://www.itmonline.org/arts/gallnuts.htm (Last accessed: August 8, 2011.)

Eravşar, O. 2011. *Anadolu Selçuklu Kervansarayları*. Selçuklu Belediyesi Özel Koleksiyonu, Doğan Burda Dergi Yayıncılık ve Pazarlama, İstanbul, Turkey.

Federico, G. 1997. *An Economic History of the Silk Industry*. Cambridge University Press, Cambridge, U.K.

Foltz, R.C. 1998. *Mughal India and Central Asia*. Oxford University Press, Oxford, U.K.

Foltz, R.C. 1999. *Religions of the Silk Road: Overland Trade and Cultural Exchange from Antiquity to the Fifteenth Century*. St. Martins Press, New York.

Frye, R.N. 1996. *The Heritage of Central Asia: From Antiquity to the Turkish Expansion*. Markus Wiener Publishers, Princeton, NJ.

Fung, V.C., W.K. Fung, and Y. Wind. 2007. *Competing in a Flat World: Building Enterprises for a Borderless World*. Pearson Prentice Hall, Upper Saddle River, NJ.

Gaastra, F.S. 1999. Competition or collaboration? Relations between the Dutch East India Company and Indian Merchants around 1680. In Eds. Chaudhury, S. and Morineau, M. *Merchants, Companies and Trade: Europe and Asia in the Early Modern Era*. Cambridge University Press, Cambridge, U.K.

Hansen, V. 2007. *The Impact of the Silk Road Trade on a Local Community: The Turfan Oasis, 500–800*. Available at http://www.yale.edu/history/faculty/materials/hansen-silk-road-trade.pdf (Last accessed: August 8, 2011.)

Haudrère, P. 1999. The French India Company and its trade in the 18th century. In Eds. Chaudhury, S. and Morineau, M. *Merchants, Companies and Trade: Europe and Asia in the Early Modern Era*. Cambridge University Press, Cambridge, U.K.

Hinz, W. 1969. The value of Toman in the middle ages. *Yädnäma-i irani-i Minorsky*, Tehrän, Iran.

Jacoby, D. 1997. The migration of merchants and craftsmen: A Mediterranean perspective (12th–15th century). In *Trade, Commodities and Shipping in the Medieval Mediterranean*. Variorum Collected Studies Series, U.K.

Lewis, M.E. 2009. *China's Cosmopolitan Empire: The Tang Dynasty*. Harvard University Press, Cambridge, MA.

Liu, X. 1996. *Silk and Religion: An Exploration of Material Life and the Thought of People AD 600–1200*. Oxford University Press, Oxford, U.K.

Liu, X. 2001. The silk road: Overland trade and cultural interactions in Eurasia. In Ed. M. Adams. *Agricultural and Pastoral Societies in Ancient and Classical History*. Temple University Press, Philadelphia, PA.

MacKerras, C. 2000. Uygur–Tang relations, 744–840. *Central Asian Survey*, 19(2):223–234.

Marks, R. 1998. *Tigers, Rice, Silk and Silt: Environment and Economy in Late Imperial South China*. Cambridge University Press, Cambridge, U.K.

Matthee, R.P. 1999. *The Politics of Trade in Safavid Iran, Silk for Silver 1600–1730*. Cambridge University Press, Cambridge, U.K.

McPherson, K. 2004. Staying on: Reflections on the survival of Portuguese enterprise in the Bay of Bengal and Southeast Asia from the 17th to the 18th century. In Ed. Borschberg, P. *South China and Maritime Asia 14 "Iberians in the Singapore- Melaka Area 16th to 18th Century."* Harrassowitz Verlag, Germany.

Morineau, M. 1999a. Eastern and western merchants. In Eds. Chaudhury, S. and Morineau, M. *Merchants, Companies and Trade: Europe and Asia in the Early Modern Era*. Cambridge University Press, Cambridge, U.K.

Morineau, M. 1999b. The Indian challenge: 17th to 18th centuries. In Eds. Chaudhury, S. and Morineau, M. *Merchants, Companies and Trade: Europe and Asia in the Early Modern Era*. Cambridge University Press, Cambridge, U.K.

Oka, R. and M. Kusimba. 2008. The archeology of trading systems, part 1: Towards a new trade synthesis. *Journal of Archaeological Research*, 16(4):339–395.

Palat, R.A. and I. Wallerstein. 1999. Of what world system was pre-1500 India a part? In Eds. Chaudhury, S. and Morineau, M. *Merchants, Companies and Trade: Europe and Asia in the Early Modern Era*. Cambridge University Press, Cambridge, U.K.

Phillips, C. 1990. The growth and composition of trade in the Iberian Empires, 1450–1750. In Ed. Tracy, J.D. *The Rise of Merchant Empires: Long Distance Trade in the Early Modern World, 1350–1750*. Cambridge University Press, Cambridge, U.K.

Prakash, O. 1999. The Portuguese and the Dutch in Asian maritime trade. In Eds. Chaudhury, S. and Morineau, M. *Merchants, Companies and Trade: Europe and Asia in the Early Modern Era*. Cambridge University Press, Cambridge, U.K.

Rossabi, M. 1990. Decline of the central Asian caravan trade. In Ed. Tracy, J.D. *The Rise of Merchant Empires: Long Distance Trade in the Early Modern World, 1350–1750*. Cambridge University Press, Cambridge, U.K.

Sinor, D. 1982. Interpreters in medieval inner Asia appeared in Asian and African studies. *Journal of the Israel Oriental Studies*, 16. Haifa, Collected in *Studies in Medieval Inner Asia*. Variorum Collected Studies Series, U.K.

Sinor, D. 1997. Languages and cultural interchange along the Silk Roads. Collected in *Studies in Medieval Inner Asia*. Variorum Collected Studies Series, U.K.

Spooner, F.C. 1972. *The International Economy and Monetary Movements in France, 1493–1725*. Harvard University Press, Cambridge, MA.

Steensgaard, N. 1990. The growth and composition of the long-distance trade of England and Dutch republic before 1750. In Ed. Tracy, J.D. *The Rise of Merchant Empires: Long Distance Trade in the Early Modern World, 1350–1750*. Cambridge University Press, Cambridge, U.K.

Steensgaard, N. 1999. The route through Quandahar: The significance of the overland trade from India to the West in the 17th century. In Eds. Chaudhury, S. and Morineau, M. *Merchants, Companies and Trade: Europe and Asia in the Early Modern Era*. Cambridge University Press, Cambridge, U.K.

Subrahmanyam, S. 1997. The Coromandel Malacca trade in the 16th century: A study of its evolving structure. In Ed. Prakash, O. *European Commercial Expansion in Early Modern Asia*, Variorum Collected Studies Series, U.K.

Turan, O. 2009. *Selçuklular Tarihi ve Türk-İslam Medeniyeti*. Ötüken Neşriyat, Beyoğlu, İstanbul, Turkey.

Veinstein, G. 1999. Commercial relations between India and Ottoman Empire. In Eds. Chauffeur, S. and Morineau, M. *Merchants, Companies and Trade: Europe and Asia in the Early Modern Era*. Cambridge University Press, Cambridge, U.K.

Weisbrod, G. 2008. Models to predict the economic development impact of transportation projects: Historical experience and new applications. *The Annals of Regional Science*, 42(3):519–543.

Whitfield, S. 1999. *Life along the Silk Road*. University of California Press, Berkeley and Los Angeles, CA.

Wood, F. 2004. *The Silk Road: Two Thousand Years in the Heart of Asia*. University of California Press, Berkeley and Los Angeles, CA.

Yavuz, A.T. 1997. The concepts that shape Anatolian Seljuq caravanserais. In Ed. Gülru Necipoglu. In *Muqarnas XIV: An Annual on the Visual Culture of the Islamic World*. E.J. Brill, Leiden, the Netherlands, pp. 80–95.

Chapter 2

The Silk Road Linking Artisans in India to Designers in Italy and World Markets

Ananth V. Iyer and Andrea Lenterna

Contents

2.1 Indian Artisans and Supply Chains

Roll back history a century or two, and the world had very different global supply chains. The palace of the King of Hyderabad, for example, even today shows carpets from Belgium; cut glass, pianos, and organs from Germany; tea sets from England; and so on. The supply chain went from West to East, with money or gold coins flowing from East to West. But the royalty in India also had access to their own stable of artisans—also called karigars. The karigar families handed their craft from generation to generation and spent their lives at the service of royalty. Training was through apprenticeship with an expert. The goal was to create finery for royalty with uniqueness and style dictated by the mores of the times.

As India moved to become a democracy, these tradesmen remained in society, albeit in the background. Their sense of style was born from centuries of collective wisdom and was focused on evoking the royalty of the past. Their choices of fine silks, threads, and precious and semiprecious stones seemed to belong to an era gone by. But the purpose of this chapter is to present a new supply chain—one that leverages the artisan of the East with markets in the West. We claim that these connections, facilitated by intermediaries who straddle the bridge, are a vital conduit for supply chain flows along the new Silk Road from East to West.

We will provide a specific case example that identifies the flows across the supply chain for one product—embroidered gowns. This chapter is based on a case study done by one of the coauthors, Andrea Lenterna, and his description of the details of this company (Lenterna, 2008). The company is headed by a person born in India but currently an Italian living in Italy. It was founded initially to enable outsourced manufacturing of high-end apparel in India. But soon the focus turned to enabling Western designers to get access to unique embroidery patterns created in India. How could this company play a role between the designers in Italy and the karigars in India?

The karigars are of two types—sampling karigars and production karigars. The sampling karigar is a person considered to be an artist—one who comes up with original designs that are steeped in tradition. Given their training, their goal is to continually generate possible new designs though these designs are influenced by broad themes described by their clientele. The production karigar is a person who generates multiple copies of a design that is already provided, their focus is on consistency across units and fidelity to the original design that is provided. Getting these karigars to work for you and deliver results on time is a challenging task—one that typically involves an agent in India who will get things done.

But getting work contracted and done in India, like in many developing countries, is challenging because prices charged vary with the client. Just consider a Western tourist and the prices he/she would be charged to get things done versus a local negotiator. Consider the yelling and arguing, all part of the process of interaction, to get things done. But an agent, capable of getting things done locally, would need an Italian counterpart who would understand their role and, in turn, coordinate with the wishes of the designer. The supply chain thus looks like Figure 2.1.

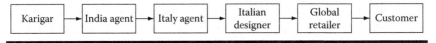

Figure 2.1   A supply chain showing entities involved.

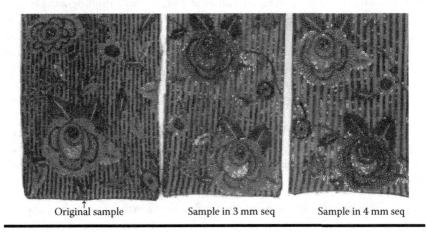

Original sample   Sample in 3 mm seq   Sample in 4 mm seq

Figure 2.2   Samples in three variants. Samples differ in the size of the stones used.

We will use fictitious names but the details will reflect current practice. Let's call the Italian agent company, Arfa. The time line for events is as follows:

Day 0: Arfa contacts designers to learn about the trends of the upcoming season. In this initial conversation, designers provide some guidance as to the type of designs and patterns they intend to use for the new collection. Instructions are usually vague and might be as simple as "We wish to see designs from the 1980s" or "Flowers will be the theme for the new collection." It is Arfa's responsibility to interpret these guidelines and provide relevant material to the designer. After all, the complexity of embroideries does not allow for descriptions to be too specific. Embroideries usually differ in their style, materials used (stones, beads, palettes, crystals, or cloth), their combination, shades of color, and manufacturing technique (refer to Figure 2.2 for an example).

Day 1: Based on what the designer asks for, Arfa needs to create a catalog of offerings. At this stage, therefore, "embroidery" means a selection of samples, usually 30 cm × 40 cm, that the designer wants to see, touch, and feel. The CEO of Arfa assembles the catalog by choosing samples from an internal library. The library contains swatches of all kinds and allows Arfa to be responsive to the needs of the designer. Therefore, it is important that the library is extensive and is in sync with the current fashion trends. When that is not the case, new samples are commissioned to the supplier and imported from India within 2 days. These samples are created by highly skilled craftsmen known as "sampling karigars." They are embroiderers whose sole task is to create new swatches. Newness is measured along the dimensions of design, materials, and manufacturing techniques. The creativity required to innovate comes at a premium. Sampling karigars are usually more expensive than production karigars (used for manufacturing) and need to be paid on a weekly rather than monthly schedule. Arfa's supplier was not always able to meet these conditions and this impacted the ability to retain sampling karigars. The result was that Arfa was never guaranteed that new samples would be available in time for the upcoming fashion season.

Day 3: An initial meeting is set up between Arfa and the designer. The purpose of the meeting is for Arfa to present the catalog of samples assembled according to the designer's specifications. Usually, Arfa's CEO presents a collection of 100–150 samples. Designers further screen the samples by selecting and holding 30–40 from this set.

Day 20: The designer holds the samples for 2–3 weeks in order to study them and assess their match with the current trends. The result of this process is a further selection that narrows down the final choice to two or three designs. At this point, a new meeting is set up with Arfa to communicate which samples will actually be used for the new collection. It is only at this stage that the selected two to three samples transfer ownership from Arfa to the designer while the remainders are returned. In the meeting, the designer not only states which samples will enter the collection but also communicates what the samples will be used for. To do so, the designer presents two sets of drawings. The first set includes nontechnical sketches showing what the final product should look like. These drawings present the overall look of the design and layout of the location of the embroidery. The second set, on the other hand, comprises technical drawings and paper models showing the shape and dimensions of the cloth as well as the specific locations on the cloth where the embroidery needs to appear. At this stage, therefore, "embroidery" assumes a new meaning: not 30 cm × 40 cm swatches anymore but embroidered components of a dress. This change of definition is reflected by a new type of production process. While swatches are usually created by the more skillful sampling karigars, the components of a dress are created by the cheaper production karigars (usually 15%–20% cheaper). This happens because replication is easier than innovation and, although production karigars are responsible for the quality of the final product, they simply have to duplicate an existing design.

Day 21: Arfa contacts the supplier in India to communicate which samples have been selected. Arfa also turns the information received from the designer directly to the supplier. Information flows through several channels. Paper models are sent via couriers and usually take 2 days to reach the supplier in India. Technical and nontechnical drawings are scanned and sent via email. General instructions and clarifications are communicated via phone. All this information allows the supplier to assess the type and quantity of materials needed to create a first prototype dress. The procurement process that takes place at this stage is more of an art than a structured approach. The various materials are purchased in a market comprised of a myriad of small shops in India where personal relations are very important and a handshake is sufficient for any deal to go through. But this requires trusted parties to be involved, hence the need for an Indian agent to consummate the deal and get the products made.

Day 25: The supplier can create the various components of a dress within 2 days. Arfa receives the various parts 2 days later via courier, inspects the quality of the product, and turns it to the designers. The designers go through a second quality inspection before turning the merchandise to their internal stylists. What the stylists receive are a set of embroidered pieces that have to be cut to their final shapes and stitched together. Only at this point the dress materializes and the brand label is attached. The prototype dress is ready for the fashion show.

Day 26: The fashion show is the first step in the process where the designers can assess the demand for any one dress. Dresses are luxurious, complex, and unique and it is virtually impossible to predict how successful any one model will be. This is why production is performed under a make-to-order process and production schedules are put in place only after the fashion show, when customers express their interest but don't commit yet to purchasing the dress. A sense of demand is also required by the supplier to hold any inventory of raw materials, which are kept at a minimum of 50% of the order size.

Day 27: After the fashion show, a new meeting is scheduled. The purpose of this meeting is to investigate ways to modify the prototype dress in order to reduce its final price. Price, in fact, is rarely a concern prior to the fashion show. There are at least two reasons for this. First, the designer is willing to sacrifice price in lieu of superb quality on the runway. Second, there is often very little time for any price negotiation to take place. In order to save time and ensure that the dress will be ready for the show, the designer might accept a higher price being well aware that it might be too high for any customer to bear. Because of this, the prototype dress often undergoes changes that reduce the final price to the consumer. Dresses are usually reshaped by two separate forces—the designer and Arfa. The designer modifies the structural elements of the dress in order to reduce complexity and, thus, cost. Arfa, on the other hand, proposes alternative materials or production techniques that lower the manufacturing costs of the embroidery (refer to Figure 2.2 to see variants of the same design).

Day 34: A new meeting is set up between Arfa and the designers. The purpose of the meeting is to finalize the changes to be made on the dress and to agree on a final production price. By the time the meeting takes place, Arfa has a sense of how much each alternative will cost and has new swatches ready for the designer. In this way the designer can assess the tradeoff between cost and quality and readily decide which one is the best alternative. At this stage, the designer is also able to communicate a final price to the customer and receive actual orders for each dress. Soon after the meeting the designer is able to place an order with Arfa. The order specifies the quantity to be produced and the expected delivery time.

Day 44–49: Once the designer places an order with Arfa, Arfa turns the order to the supplier.

The order specifies which alternative has been selected by the designer, the quantity to be produced, the expected delivery time and any last minute adjustments that the designer wants on the dress. Such changes are communicated to the supplier via telephone or email, new design instructions are scanned and sent as attachments, and, if there are new paper models, they are sent via courier. The production process usually takes between 10 and 15 days, after which the whole production batch is sent back to Arfa. Arfa receives the completed products between days 44 and 49, performs a final quality inspection and turns the whole order back to the designer. At this point Arfa exits from the chain of events. What happens next is all within the designer's facilities and consists of cutting the various components of the dress into the proper shapes and stitching them together. The final step is attaching labels before the dresses are delivered to the customers.

To understand the extent of changes between the design shown at the runway and the design that is produced, refer to an article by Holmes and Dodes (2011) in the Wall Street Journal. One designer describes changes in material (leather to crinkled satin), length of the dress, number of bows, etc., that the article cites as decreasing prices from USD 798 to USD 398. The price had to be adjusted to respond to retailer concerns about customer willingness to pay and associated sales volumes. A similar negotiation ensues between Arfa and the designer. The main difference is that Arfa, in turn, has to work with the karigars in India to adjust the embroidery to generate possible designs.

2.2 Different Way of Doing Business

At no point in the evolution of the business does Arfa sign any contract, other than letters of credit, with either designers or its supplier. Relationships and trust are the key that allow Arfa to operate in the fashion industry. Creating a dress for a fashion show is a frantic task. Everything has to happen within a few days and last minute changes are the norm. In addition to this, there are virtually an infinite number of things that can go wrong. The dress could miss the deadline, the color of the stones could be slightly off, and design patterns could not perfectly meet specifications. Structuring relationships through contracts are simply not feasible. A lot of what happens within the industry, therefore, has to happen on the basis of trust, creating one of the greatest paradoxes ever. The industry could not solely use contracts if it were to function, yet designers had to ensure the uniqueness of their models. Two designers just could not afford to come out with the same design if their brand was meant to stand for luxury and exclusivity.

Just as no contracts were in place between Arfa and the various designers, they were also not a part of the business between Arfa and the Indian agent. Despite the objective inconvenience of setting up contracts, operating on the good faith of others was how agents conducted business in India. The heavy reliance on trust does not necessarily mean that everything runs smoothly. In one instance, a designer turned to one of Arfa's competitors only to find out later that the embroideries he commissioned had already been used by others. Although the dresses were ready for the runway, he decided to burn all of them just before the fashion show!

Letters of credit are the usual means of payment. Under this arrangement, Arfa has to anticipate a payment to the Indian agent before even receiving the merchandise. The funds are supposed to cover the expenses of raw material and are necessary for the job to get started. The Indian agent constantly claimed not to have sufficient funds. It was only after the completion of the work that the Indian agent would cover his liability with Arfa. The initial payment would be subtracted from the final invoice and Arfa would only have to pay for the difference. This procedure, however, created a wrong set of incentives for the agent in India. In fact, once the first portion of the payment was received, the Indian agent had little motivation to remain on schedule and on-time delivery of products required constant pressure from Arfa. In addition to this, the anticipated payments for the raw material also created an opportunity for the agent to mask his real costs. Arfa was well aware of these limitations but this was the way the relationship had been set up initially, when prices were definitely competitive, and, now, it was hard to change the "way of doing business."

2.3 Value Proposition

The role of Arfa as an intermediary had been evolving over time. In the early 1990s, the value proposition was very simple and exploited frictional forces between the Italian and Indian market. Despite the beauty and cost savings of Indian embroideries, no designer had experience with outsourcing production to India and significant barriers existed due to cultural and linguistic differences. Moreover, confidentiality remained a critical issue since designers were worried about intellectual property leakages resulting from outsourcing production to India. In such an environment, the Italian agent Arfa would be the bridge between the two worlds. An agent who knew what it meant to work both in India and in Europe and was able to knock out any linguistic barriers would thus play a key role in enabling transactions. On the one hand, Arfa was at a disadvantage because it did not have direct control over the production process in India. On the other hand, however, designers had a readily available interlocutor with whom to talk personally. This allowed Arfa to win the designer's trust and become the classical middle man importing embroideries from India, taking care of any logistic issues and delivering the final product for a price comprehensive of production costs, transportation costs, and import taxes.

As time passed by, however, the industry observed several changes that could undermine Arfa's value proposition. The Indian population was becoming increasingly more educated and the language barriers that existed in the early 1990s were disappearing. Meanwhile, the rising competition in the airline industry exerted a downward pressure on airfares, thus increasing the global mobility. The combination of these two factors allowed any Indian manufacturer to come to Italy and deal directly with designers, thus eliminating the layer of the supply chain occupied by Arfa with consequential savings to be shared between designers and manufacturers. For Arfa to survive, the business proposition had to adjust to the changing environment.

One way to battle in the competitive arena was to leverage what, over the years, had become an additional source of value for designers, namely the library of sample embroideries. By 2008, the library contained roughly 2000 samples that provided Arfa with a pool of readily available materials to show to the designers. The presence of an internal library was definitely valued by stylists but Arfa was quite sure it was not reaping the full benefits of this resource, imputing part of the responsibility on the supplier.

A closer look at how the samples were created can explain the magnitude of the problem. Ideally, the generation of new samples was to take a two-phase approach. The first phase was the creative one where the goal was to generate ideas for new designs. In order to fulfill this goal, Arfa relied on several inputs. Arfa would create designs internally, buy them from external professionals or even get inspired from paintings and pieces of artwork. Despite the source used, Arfa did not have a structured approach to update its library. New trends would be anticipated by specialized companies that would share their findings with the major designers only. All that Arfa could do, therefore, was to second guess what the trends were going to be by talking to designers. The second phase, on the other hand, consisted on turning ideas into actual embroideries and relied on the mastery of the sampling karigars. This was the phase in which new production techniques and new materials were to be tested. This two phase approach, however, often remained a pure idealization of a process that should have happened but that rarely did. Because the Indian agent did not pay the sampling karigars on a timely fashion, the designs that Arfa sent to India rarely were recognized. Instead, it was the karigar himself that would create new designs and implement them. This did not allow Arfa to have samples in sync with the emerging fashion trends and, even if its designs were to be used, Arfa would receive them only after the fashion season had ended.

2.4 2008 Collection

In January 2008, Arfa was getting ready for the Spring/Summer collection. As usual, the company contacted various designers to receive inputs about the new trends of the year and obtain guidelines about the type of patterns, designs and materials that each designer was looking for in the coming season. After several years in the industry, Arfa had developed an internal library of sample embroideries categorized by style, production processes, and material. This library allowed Arfa to quickly sort through the samples and select those that matched the designers' requirements. For the upcoming meeting, Arfa had selected 150 samples.

For the 2008 collection, the client asked for samples in red, containing flowery patterns. Out of the 150 samples that Arfa preselected for the meeting, the styling department chose 10. It was customary in the industry to leave the samples with the designer if asked to do so; at this stage, the samples remained the property of Arfa. The designer held the samples to further study them and determine if they conformed to the trends of the upcoming season. The investigation process could take as long as 1 month after which the designer would contact Arfa and communicate the final decision. The practice of holding the samples was also utilized as a preventive measure by the designer. With the sample not being readily available to Arfa, the same design could not be sold to multiple designers. Exclusivity of the design is crucial in the fashion industry.

The investigation process would usually result in some of the samples being rejected. Out of the 10 samples initially selected by the client, 4 were chosen while the remaining 6 were returned. At this point, the chosen samples transferred ownership from Arfa to the designer who owned all rights on the design. The six samples that had been rejected, on the other hand, were still owned by Arfa who was free to show them to other designers. The selling fashion season usually lasted 3 weeks and, if the samples were returned fast enough, Arfa could show them to other designers. If, on the other hand, the selling season had passed by, Arfa would take the samples back to the office and store them in the company library. The same samples could be used in future seasons in an industry where trends come and go in a cyclical pattern.

One of the samples chosen by the client was the F511 R4 (refer to Figure 2.2). Based on the sample's design, the stylist designed a complete dress to be shown at the fashion show in February. The dress comprised of a fully embroidered skirt and a top that were to be completed no later than January 25. Although the designer had a clear picture of how the dress was supposed to look like, Arfa was not responsible for the completion of the whole dress. Instead, it would be supplied with paper models spelling out instructions for what the part was to look like. The Indian agent would receive the paper models directly from Arfa by air shipment. The models would instruct the karigars about the shape of the part and where exactly the embroidery was supposed to appear. At no point, however, were the karigars supposed to take care of the final cutting and stitching of the complete dress. Those tasks were performed in house by the client who also maintained the right to attach its own label. Most of the times, this prevented the karigars from knowing whose dress they were working for even if they were well aware that the final customer was going to be one of the top names in the fashion industry.

The dress successfully made it to the fashion show but several changes were to occur before final production took place. Pricing was always an issue in the luxury fashion business but it was often addressed only after the fashion show. Designers were not concerned about pricing for the prototype dress to be shown on the runway. What mattered, at that stage, was that the dress looked as beautiful as possible under the flashing lights of photographers. This was the dress that was going to appear in the most respected fashion magazines around the globe and designers were willing to sacrifice low prices for superior quality and timely delivery.

The same was not true of the dress to go into production. Designers were well aware that the commercial version of the dress could sacrifice superior materials and execution complexity in lieu of lower price points that would make the dress more accessible to the customer base. This was also the case in this instance and the dress was to undergo some changes. Arfa had been asked to identify the cost drivers for the specific design and provide alternatives that would bring costs down. It was later determined that the size of the stones used were a significant factor. The smaller the size the more labor intensive the process would have been, thus commanding higher prices. The prototype dress used 2 mm stones, imposing a final price to the stylist of 1150€. By using stones of 3 and 4 mm, Arfa was able to quote prices as low as 750 and 445€, respectively. The client accepted the 3 mm version while simultaneously presenting some changes to be made to the dress as a whole before entering final production. The changes were made to reduce the complexity of the dress and achieve a lower final price of 674. Only after an agreement on the final production price was reached, did manufacturing begin.

2.5 Flow of Goods from East to West

The supply chain described in the case shows the benefits of splitting up a supply chain into many different pieces—each permitting more effective contracting and delivery. The anonymity of subsequent steps afforded by an agent justifies the agent's role in preventing intellectual property leakage—which for apparel, is crucial in holding value and defining the designer's fashion sense. By withholding identity of the designer until the final stage, albeit with associated higher finishing costs, the supply chain structure displays a tradeoff between efficiency (which is perhaps lower) and survival in a competitive atmosphere where copying a design is easy. A more important aspect of the supply chain is the role of intermediaries in enabling collaboration between the sense of taste by the Indian karigars, developed over years of design creation, and the sense of Western taste that is determined by the Italian designer. If supply chains end up requiring intermediaries for leveraging of global capability, then that is a great lesson for today's global supply chain managers.

References

Holmes, E. and R. Dodes. 2011. Materials girls: Designers trim hemlines, costs. *Wall Street Journal*. February 17, 2011, D1.

Lenterna, A. Arfa: Where is the value. MBA independent study report, supervised by Professor Ananth V. Iyer, Krannert School of Management, Purdue University, West Lafayette, IN, 2008.

Chapter 3

Logistics Management Insights from the Silk Road Geography

Ercan Korkut

Contents

3.1 Introduction

In this chapter, I aim to share my hands-on lessons and insights over a period of 10 years spent in the logistics management of the countries in this geography with special emphasis on the changes and trends by linking to the legislations, regulations, and infrastructure. I will also underline the aspects deserving special attention when establishing a supply chain in this geography. Though my experiences are telecommunication centric, I think that my hands-on lessons may still provide useful guidance to other sectors as well.

3.2 Three BRIC Countries out of Four Are on the Silk Road: Can It Be a Coincidence?

Today, almost everyone in business is aware of the BRIC countries namely Brazil, Russia, India, and China. They managed to sustain their growth even as crises shook the globe. They are forecasted to become the four most dominant economies of the globe by the middle of the twenty-first century.* Can it be explained only as a coincidence?

It is impossible to explain it as a coincidence. Behind this success, there are strong visions covering long-run strategies being implemented by determination to get such visions.

The Silk Road has been regaining its importance from past centuries as well as its influential power in the arena of world trade. However, it no longer consists of caravans passing along the Silk Road, but of digital signals and software through the fiber optical data lines, services through the call centers, and natural gas and oil passing through the pipe lines. The contribution of China and India to the development of the Silk Road geography is not only due to their supply power, but also due to their demand power. It is true that China is a center of production for global brands with its cheap labor force, and India is a center of services and software development with its educated information technology force. On the other hand, they are also attractive markets with their total population already in excess of 2.5 billion.†

3.3 Observations over the Last Decade: What Has Changed?

Until a few years ago, when we offered our customers telecom equipment manufactured in China, the answer from the customers in Kazakhstan, Turkey, Ukraine, or Iran, regardless of where they were based, was a sharp "No." There was a bitter perception on the low quality of the China offerings and as a result there was a strong negative reaction. This negative perception was valid not only for the "no name" production in China, but also for the production by third parties under the license of the global brands and even for the production by the global giants at their own factories in China.

Some customers' reactions went even further in a way that they were adding contractual clauses to get the seller's commitment not to use China made

---

* http://en.wikipedia.org/wiki/BRIC (The BRIC thesis, Goldman Sachs).
† Population as of 2010·09·26: China 1,331,694,827, India 1,176,966,740, available at http://www.xist.org

products. As a result, the factories of the global brands established in China were underutilized for many years.

Today, the picture is exactly the opposite. The customers have been demanding even the products manufactured under the license and technology of the Chinese. Two Chinese telecommunication vendors have already become major players in the industry. There are a couple of reasons behind this incredible transition that took place in less than a decade. The most important one among those reasons is that Chinese firms caught up with their Western rivals in terms of quality while still managing to keep their costs low. In parallel, the financial crises forced the operators to be cost centric more than ever because of the continuously reducing revenues per subscriber as well as the investment pressures caused by the need to renew their networks every 2–3 years in order to keep up with the developing technologies (second-generation, third-generation, fourth-generation long-term evaluation). The mobile operators do not see a rationale to invest in the best class quality networks that will have to be renewed in 2–3 years time. A moderate quality level of network meeting the expectations of the operators regarding capital expenditure (CAPEX) and operational expenditure (OPEX) is found to be more than satisfactory. Therefore, Chinese telecom vendors who are capable of meeting these expectations with minimum CAPEXs face a demand boom. The same trend is also applicable in markets other than telecommunications.

In response to the Chinese rivals, the Western vendors initiated the following new strategy to change their market image from "high class but expensive" to "cheap but also high quality." They applied strategies to prove that quality and economics can improve together. To serve this purpose, they moved their production to countries such as China, India, Malaysia, Indonesia, and Thailand, where the labor force is cheaper, by increasing their established production capacities in such countries. In this way, Western vendors responded to their Chinese competitors and improved their quality and gained competitive advantage by getting access to low-cost resources. It is no longer valid to say in the telecommunication industry that the price level is a competitive advantage for Chinese telecom vendors. Knowing that capital has been accumulating in both China and India, I expect that both these countries will aim to become global brands by mergers and acquisitions in the West. China may be able to act faster because of the USD 2.5 trillion cash in its treasury.

In parallel to the shift of production premises to the Silk Road geography, the regional distribution centers (hubs) started to increase. Nokia Siemens Networks has two hubs and five production facilities in the geography. As the standard production costs become lower, the inventory carrying costs decrease. This encourages the use of low cost but environmentally friendly transportation modes more intensively. It means that the portion of the seaway and railway in the transportation market will increase its acceleration.

Another flashing development in the telecommunication industry is the increasing proportion and importance of the software. Developed software and internet infrastructure enable the loading of the software and upgrades on the hardware remotely without visiting the site. The capacities of the hardware can be increased; the failures and outages can be intervened remotely even from thousands of miles away. Luckily, the internet infrastructure of the countries in this geography is sufficiently developed to allow downloading of data storable only by tens of CDs. Nevertheless, the foreign trade legislations of such countries are not adapted to virtual trade over the internet. Hence it is mandatory, especially in the ex-Soviet countries, to get the same software—already downloaded from the Internet—once more, but this time physically on a media (i.e., data tape, CD) through the customs in order to prove the transaction by using conventional methods. For this reason the supply chains in some countries of the Silk Road still need to be modeled in order to support multiple ways of distribution.

3.4 Hands-On Lessons and Insights

In this section, I will present important incidents with embedded lessons learned in several countries. The legislative noncompatibility between many ex-Soviet countries and European Union countries may be the first issue among others, to pay attention to. Since the beginning of 2009, EU countries are obliged to use e-customs that practically means no paper is used anymore other than the export accompany document (EAD); the customs at the border control everything in connection with values, tariff codes, and the exporter from computer screens. The value declared in the EAD aims to collect only the statistics. The customs of some countries such as Ukraine, Belorussia, and Russia want to cross-check the correctness of the import value declared by comparing the import invoice, the contract between the seller and the buyer, and also the export value declared at the export customs. This is the point where a problem may start. Let us assume that a seller registered in EU sells a software of a license to a buyer based in Ukraine with a final price of 400K€ after a 60% discount over the original price of 1M€ set forth on the frame agreement some time ago. Software is intangible according to EU legislation and only the carrier media value (CD) is declared in the EAD—that is only 5€. The customs officer in Ukraine customs has now three different values: 5€ on the export declaration form, 400,000€ on the customer invoice declared to the Ukrainian customs, and 1,000,000€ stated on the frame contract. The customs officer gets confused and to be on the safe side against any tax evasion, forces the importer to pay tax based on 1M€, which is not fair because the real value of the transaction is only 400K€.

Triangular trade* is another similar example. Triangular trade is a way of foreign trading among three parties: seller, buyer, and a third party usually a manufacturer. Instead of the seller, the third party ships the goods directly to the buyer while the seller invoices the buyer. In other words, the country where the invoice is issued is different than the country from where the goods are shipped. However, such triangular trade causes problems similar to the one mentioned earlier. Especially, in case the buyer and the third party are both registered in the European Union, the third party declares to the export customs the selling price to the seller. However, the buyer declares to the import customs the buying price from the seller. Therefore, the customs officer in Belorussia cannot match the export declaration and the import declaration values, which are naturally different. This triggers a suspicion on tax evasion. Should the seller and the third party be two subsidiaries of the same parent company having transfer pricing (intra-company invoicing), such suspicions become greater.

Such cases due to the inability to adapt the country legislations to the world standards resulted in incidents where the goods were captured by the state and taken to the court. They are barriers not only to the development of the supply chain but also to the development of commerce. It is imperative to synchronize and standardize the legislations in those countries, but that task is under the initiatives and jurisdiction of the governments. Until that happens, companies should get acquainted with the legislations of the countries in this geography. As an example, ATA-Carnet† is a foreign export tool used to ease the temporary movement of the goods and samples for demonstration or trial purposes among the countries without paying duty or tax with almost no customs formalities. It is accepted in 75 countries as of September, 2010.‡ Kazakhstan, Azerbaijan, Iran, Kyrgyzstan, Turkmenistan, Tajikistan, Uzbekistan, and Armenia do not accept ATA-Carnet application. Hence, when one needs to demonstrate a commercial sample to its customers in those countries, one has to fulfill tremendous amount of temporary importation formalities and pay duties at import customs.

Another hands-on lesson was learned in Uzbekistan. You cannot move even a stone from one address to another without a contract signed between buyer and seller. Especially for foreign trade, the contract needs to be approved by the Central Bank. Otherwise, the currency cannot be transferred abroad. It may take sometimes months to get such approvals. It is necessary to list each and every single item to be traded including the subitems in the contract. A negligence of even a subitem may result in inability to import it. This causes serious problems mainly for partial shipments. In case the contract lists the main system

---

* http://www.export911.com/e911/export/triTrade.htm
† http://en.wikipedia.org/wiki/ATA_Carnet
‡ http://www.uscib.org/index.asp?DocumentID=1843

instead of the subitems making the system and if any of the subitems cannot be shipped together with the main system due to supply problems, you will have two options—either to delay the shipment of the whole system until the missing subitem is completed or getting a separate import permission for the missing item to be replenished later. Therefore, companies should prepare itemized contractual list and always remember to plan for the time to get the contract approval from the Central Bank.

You can do the domestic transportation of the goods only by using the trucks that you contracted. This limits the supply of trucks and hence limits your agility in urgent deliveries. A global logistics company with which we partnered providing domestic logistic services such as warehousing and inland transportation in Uzbekistan was unaware of such condition mandating a contract with every truck. They became aware of such a condition and stopped all our inland transportations until they contracted with the trucks. It had negative impact on our project rollout due to the unexpected interruption of site deliveries. This proved once more the importance and even vitality of a deep analysis on the local needs and legislations prior to a new market/country entry. A quick and dirty entry without a proper analysis and preparation for the sake of speedy ramp up may cause unexpected bad surprises that are costly. Check points prior to a country entry as detailed in Section 3.5 may help to prevent the occurrence of such risks.

The political and economic volatility of the countries in the Silk Road geography necessitates capital owners to have capability of quick country entry as well as quick exit. It is mandatory to have an agile supply chain to benefit from the competitive advantage of being the first to deliver to the customer. What is meant by being agile is the ability to ramp up and down the operations in a short period of time. Ability of quick ramp up and down is important especially in project businesses. Light hubs feeding only a couple of countries with simple but reliable inventory management systems are easy to model and implement instead of sophisticated ERP systems.

Another contributor to the agility is outsourcing the supply chain operations to a reputable logistics partner. Companies can outsource all logistics operations except modeling the supply chain that should be perceived as a core task. Global companies prefer to work with global logistics partners with whom they sign a global frame contract. This saves time for the additional contract negotiations with the local partners and mitigates extra risk and additional local costs of insurance and liabilities. However, global logistics companies, which are managed to be localized, have challenges in meeting the market expectations. Contrary to their messages to the market, they do not authorize their local organizations (decentralization) and this causes delays in decision making and satisfying customer demands. Additionally, the overhead coming from the headquarters or regional organizations make the global logistics brands 30%–40% more expensive than the local logistics companies.

In this geography, the logistics companies that can manage to be GLOCAL (global and local) will gain competitive advantage. The logistics companies that know the local facts with the ability to operate quickly like a local company, as well as giving the trust and quality of a global brand will break sales records.

Admitted insurance* means that the insurer is registered in the same country that the insured risk is domiciled. Non-admitted means that the insurer is registered in another country outside of risk domicile. Countries such as Turkey, Azerbaijan, Belorussia, India, Indonesia, Iran, Kazakhstan, Malaysia, Moldavia, Pakistan, Russia, and Uzbekistan do not accept non-admitted insurance. For instance, if you buy your warehouse insurance policy from an insurer not licensed in Uzbekistan where your warehouse is located, in case of a fire at your warehouse, you will be fined due to noncompliance to the local insurance legislations. Also you may have difficulties in claim payments done locally and the additional taxes transferring the claim payment. It is correct that the master policies of the non-admitted insurances are easy to issue and cheaper, yet in case the risk materializes, the consequences would be dear. Hence, I recommend to the global firms to buy local polices from the local agents of their global insurers issuing the master insurance programs. Otherwise, buying a stand-alone local policy is inevitable. In any case, global firms should consider the extra cost of local insurance in aforementioned countries where non-admitted insurance is not valid.

There is a customs union established among Russia, Belorussia, and Kazakhstan as of January 1, 2010. Because the details of the operational instructions were not communicated early enough and the officers were not trained properly, the new legislations caused delays at customs transactions. This union leads to changes in the customs tax rates. In order to be protected from extra costs such unexpected changes may cause, I recommend sellers to prefer the international commercial term (Inco term, 2000) "DDU" (Delivery Duty Unpaid)† where the customs clearance, hence duty and taxes are under the buyer/importer's responsibility. In case it is not possible due to the competition, and the Inco term DDP (Delivery Duty Paid where the seller is responsible from customs clearance, hence the duty and taxes) is agreed, still the seller can add a contractual clause that enables the seller to reflect such changes in duty

---

* See, for example, the article titled "Global implications of admitted, non-admitted and self-insurance" by Donn Pfluger-Murray and Jason Taylor, available at http://findarticles.com/p/articles/mi_qa5332/is_10_52/ai_n29213273/

† Inco term DDU (Delivery Duty Unpaid) of Inco terms 2000 is replaced by DAP (Delivered at Place) in Inco terms 2010.

and taxation. Should this union be successful, it is expected that the union will enlarge with the participation of other countries. For the companies that have intensive commerce with those countries, this union may bring advantages if the supply chain is modeled properly. For instance, a distribution center (hub) or an assembly premise even at semi-knockdown level or both may result in faster and cheaper circulation of goods among the union members where there are no customs barriers. This may lead to competitive advantage against India, China, and other rivals from the Far East. What I recommend to global companies is to follow up closely such trade agreements in this geography to create advantages in the competition.

Another aspect deserving attention in this geography is the trade restrictions and embargos. Trade restrictions depend on the technical specification of the goods, the export and the import countries, and the end user. Iran, North Korea, and Syria are three countries out of five (others are Cuba and Sudan) that are subject to United Nations embargo in the Silk Road geography. Other than general embargo, there are also countries subject to individual validated license (IVL). Armenia, Azerbaijan, Iraq, Iran, Pakistan, Syria, Abkhazia, and Uzbekistan are examples of IVL cases. Prior to export to any of these countries, foreign ministries should be contacted regarding trade restrictions.

Mainly in the trade of information technology and telecommunication equipment, foreign ministries of the export countries demand written commitment from the end customers that the goods will not be used for military purposes or will not be sold to another party. If the end customer or the end user is a state body or military, foreign ministry of the exporting country may not approve IVL to export. IVL needs to be obtained for every single shipment. It is a process that may take 2–3 weeks. While participating in tenders in such countries that are subject to international trade restrictions, it is highly recommended that the bidder should precheck if the goods can be sold and whether the buyer/end user are restricted by the sanctions of the United Nations, the United States, or the export countries where the goods would be delivered from. This precheck needs to be done for every single item. Otherwise, one may face risks of failing to fulfill the contractual commitments that may end up with loss of company credibility and significant penalties. Furthermore, the lead time to obtain the necessary export/import permissions needs to be planned. The trade restrictions are more important in trading system solutions consisting of original equipment manufacturer (OEM) products manufactured by several different producers. For instance, a system solution committed to Iran cannot be sold if the system includes HP or Cisco products due to US embargo to Iran.

Trade restrictions have also the other side of the story. In the previous paragraph, we discussed the restrictions applied to the importing countries in the region. Next topic is regarding the restrictions applied by the importing countries. There are incidents where some countries in the region changed their customs tax and duty regulations suddenly in order to protect their local industries. I recall a case wherein Iran increased the customs taxes applied to the imported antennas by multiples some years ago in order to protect the local antenna manufacturer. The local antennas were very cheap, but their quality did not meet the spectrums and caused trouble in our roll out. Similarly Uzbekistan changed the radio frequencies and this made our radio inventory useless. Tens of radios in our warehouse became obsolete. Therefore, the recommendation is to add cost and time contingencies in the risk analysis for such unexpected changes in the regulations.

In my opinion, one of the key points in the development of the supply chain along the Silk Road is the improvement of the railway network (also see Chapter 12 on railways). The modernization of the current railways as well as the improvement of the interconnection between the railway and other transportation modes will contribute to the trade volume. The railway connection of Europe and Asia over İstanbul Bosphorus will make the railway transportation possible all the way from Netherlands to China. Not only the reduced lead times and freights, but also the environmental impacts such as CO_2 emission should be considered, in prioritizing the railway transportation. A few facts may help us imagine the environmental impact of the topic. Nokia Siemens Networks transported 12 containers (400 cbm, 110 ton) from Netherlands to Uzbekistan by railway. It produced only 11 ton of CO_2 emission. If we would have used airway for this delivery, it would be exactly 353 ton of CO_2 emission. Nevertheless, despite its contribution to save our environment, railway transportation took 30 days, that is nearly 10 times slower than the airway. This also means a 30 day delay in cash collection. Therefore, investments to develop the railways in cooperation with countries on the Silk Road will encourage the companies to utilize the railway more frequently.

3.5 Country Start-Up Checklist

In this section, I will share the "To Do" items from a supply chain perspective that need to be analyzed before you start a business for the first time in a country. These are applicable not only for the market entries in the Silk Road geography, but in any country of the world. Doing your homework prior to operations is more important on the Silk Road where quite many countries' regulations are not compliant with the World Trade Organization. It is worth underlining that each company may have its own dos and don'ts according to the dynamics of its own industries. The aim is to provoke you to start thinking on the basics in modeling a supply chain in a new market entry. Like a pilot checking each and every single detail while preparing for a takeoff, each and every company planning to enter a new market needs to analyze the "To Do" list for a healthy execution.

This list can also be used for the activation of new customers in the countries that are already active. New customers can come from military or state bodies and may have different demands with potentially new liabilities. This list can also be useful in anticipating the risks in case a new product, though sold to an already operational country or customer, is subject to an embargo or a trade restriction.

The first issue to identify is whether the country or end user that your product is sold to is subject to an international embargo or a foreign trade restriction. This should be done at the proposal stage.

The banks to be used in the letter of credit need to be checked if they are in the black list. In case the seller provides the financing, the conditions of financing institutions should be checked for restrictions on the supplying country or for special document requirements.

The importing process must be agreed in advance. The buyer and seller, country-specific requirements, import permits, certifications if any, preferential country of origins, trade barriers, exemptions such as zero taxation such as EUR 1 and their sustainability, and import regulations for intangible products such as a license or software must be clarified well in advance even at the tendering phase. If demo or trial imports are needed, then the validity of the ATA-Carnet or the requirements of the temporary importation should be analyzed.

The special requirements of your customers beyond the standard documents and common practices need to be learned and the capability of your company to meet such exceptional requirements should be examined. Your customer may demand a special remark on the invoice such as showing the invoice value on both hard and local currency, not writing tariff codes, not mentioning the country of origin, not showing the software, or not splitting the discounts. Your customer may require special packaging, labeling, or transportation conditions. Such requirements may not be complying with the code of conduct of your company and may even conflict with Sarbanes Oxley Act. Your company's IT system such as SAP may not be designed to meet such demands of your customer and may force you to find work-around/manual solutions that may result in more man hours. Therefore, it is mandatory to learn about such exceptional specific customer requirements as early as possible to find out their applicability prior to your commitment to the customer. In case pre-shipment inspection is demanded, the process should be reviewed with the inspector company and the details should be agreed upon.

In addition to the aforementioned import permissions, the export control process of the country of export must be checked with a special emphasis on finding if there is a trade restriction or embargo implemented on the country of import. This check must be done for each item. The application process for the special permits if any should be learned in advance.

The capacity and the capabilities of the logistics partner that can be used in the supply chain operations should be examined. A preliminary request for information (RFI) should be made to identify potential partners earlier. Since the selection of the logistics service providers is a lengthy process, it should be started much earlier and the results should be kept in the country database with regular updates activated when the market entry decision is taken. This enables the company to respond to the tenders faster and more importantly with realistic and accurate data.

In designing the supply chain, domestic logistics services required and the key performance indicators, to measure the ability in meeting customer expectations, should be identified. What the customer demands and what our supply chain is capable of need to be compared. The improvement actions to close the gaps should be identified and executed in cooperation with the logistics partners. If necessary, corrections in the supply chain modeling should be made.

Another RFI process should be conducted to collect freight quotations for international transportation based on the estimated volumes. Especially for the airway transportation, special deals are necessary to transport oversized packages (>2 mt), or dangerous goods having magnetic or flammable characteristics. Not every carrier is capable of handling such loads. It is good to know such constraints and plan accordingly to prevent last minute off loadings.

3.6 Conclusion

The Silk Road shelters different countries discussed in this chapter from political and economic aspects. In this geography, the image of "cheap but quality" has been broadcast to the global community. Besides, the weight in the world economy has been increasing on both the supply as well as the demand side. In the meantime, lack of standardization in the legislations and abrupt changes in the regulations are the main characteristics of the countries in the Silk Road geography. Agile supply chains will gain competitive advantage in the market. To achieve agility, companies should model their supply chains by analyzing the local facts and the characteristics of the countries in detail as early as possible and should always keep their risk maps updated. A lean supply chain with minimum number of layers will also minimize the transfer of risks and potential conflicts. An agile supply chain with quick ramp up capability to start operations as well as quick ramp down flexibility to close the operations will protect or at least minimize the exposure to the volatility in the geography.

The trade trends and agreements in the geography should be followed closely. It should always be kept in mind that supply chains established in such trade unions may entitle your company to become a preferred trade partner benefiting from exemptions and incentives. Logistics partners should be selected from among those who best know the local regulations and have the strongest networking inside the state organizations. Those logistics partners who manage to be GLOCAL are both reliable as a global brand and cost effective as a local brand—they can be used as partners to outsource the contextual tasks of the supply chain to make it leaner and seamless. Starting operations or even proposing a quotation needs to be avoided without analysis of the aspects listed in the start-up checklist if you intend conducting business for the first time in this geography.

Chapter 4

Formal and Informal Financial Institutions and Entrepreneurship: The Case of Bazaars and Microfinance in Central Asia

Deniz Tura

Contents

4.1 Introduction

The objective of this chapter is to analyze and compare the relationship between the formal and informal institutions and entrepreneurships in Azerbaijan, Uzbekistan, and Turkmenistan by focusing on one of the main trade centers of each country to better understand their effects on the market building processes. Entrepreneurship takes a variety of different forms and enterprises and uses different strategies in different countries, reflecting the characteristics of the external environment for private enterprise (Welter et al., 2004). Since they gained independence, all three governments have chosen very different economic and political development models. Due to large Western oil investment and more open policies, Azerbaijan is more integrated with the market economy compared to Uzbekistan and Turkmenistan. In Uzbekistan and Turkmenistan, former communist leaders consolidated political and economic power in a neo-Soviet style by curbing the operations of Western institutions and reasserting the single leader and party control in the economy and governance (Ozcan, 2006). Turkmenistan, which possesses natural resources comparable to Azerbaijan, remained extremely isolated from the outside world due to its protective policies. Uzbekistan, with the largest population in Central Asia, is still struggling with its market economy integration (Pomfret, 2006).

The unprecedented degree of institutional change experienced by the transition countries has been largely moving into a similar direction—the switch from a system based on state planning and allocation of resources dictated by the government to a system characterized by decentralized market allocation (Aidis and Sauka, 2005). The development of enterprises of any size* is critically important for the economic development for all post-Soviet Union countries. Despite the different pace of developments in their economic and political scenes, their relations with the formal institutions remain limited and the internationalization of small businesses is relatively slow. On the other hand, their relationships with the informal institutions and networks have increased and become more complex.

In the first part of the chapter, Airport bazaar in Baku, Azerbaijan, Crossroads bazaar† in Tashkent, Uzbekistan, and Desert bazaar‡ in Ashgabat, Turkmenistan are described in detail to better illustrate the business environment. In addition, the entrepreneurs of the three respective bazaars are introduced. Later, the chapter focuses on the relationship of the formal institutions and entrepreneurs including the legal, regulatory, and financial aspects as well as micro finance and

---

* For this chapter, the enterprises are classified into groups according to the number of employees. The micro enterprises consist of 1–3 employees, the small enterprises 6–20, and the medium enterprises 21–50 employees.
† Chorsu bazaar in Uzbek.
‡ Çöl pazari in Turkmen or Talkuckha in Russian.

bazaar management with case studies. In the following section, the relationship of the informal institutions and entrepreneurs are illustrated with examples. In the concluding section, the study reveals the implications of the formal and informal relationships that the entrepreneurs have on the market building processes for the three countries.

4.2 Overview of the Bazaars and Entrepreneurs

The collapse of Soviet Union and its centralized production system led to a decentralized, ad hoc, and often chaotic trading infrastructure that paved the way for the emergence of a new set of actors, traders, bazaar owners, and government officials (Spector, 2008). The demand for goods and the need for trading were very high and the infrastructure was almost nonexistent for trading spaces. For this reason, bazaars developed very rapidly and still remain as major trading centers with the highest concentration of micro, small, and medium enterprises. Bazaars are also the birthplace of the majority of larger enterprises. Another important feature is that the majority of the activities are not based on legal contracts. They are all based on nonwritten, nonbinding agreements that make the bazaar a much more interesting and complex place to understand. In addition, the entrepreneurs in the three bazaars described are also conducting their business without any IT infrastructure and they are far behind the "information age revolution" that Western enterprises are going through. In an environment where there is no electricity, constant interruption of mobile networks, and only government-controlled TV channels as in the case of Turkmenistan, the entrepreneurs are still trying to catch up with the latest trends in the world to provide to their customers. There is no world-wide-web connection in any of the mentioned bazaars. None of the enterprises use any computerized systems to control their daily cash activities, inventory, or receivables. The entries are made manually in old style books but yet there are many multimillion dollar businesses hiding under the tents and the containers. Although their boundaries are vague and the physical environment makes it more complicated, it is suggested that the bazaars are the best place to understand the business development of the enterprises in the respective countries.

The Airport bazaar in Baku, Azerbaijan, is located just opposite of the Heydar Aliyev International Airport. The "Baku welcomes you" sign outside the airport is hiding the biggest trading center in Azerbaijan and the Caucasus from the main road. After Azerbaijan gained its independence in 1991, the first entrepreneurs went to Turkey with the first commercial flights (Yüksekar, 2003). Upon their return, they sold the goods on the green field in front of the airport. The demand was so high that they did not have to make an effort to go to the city. When the airport was extended, the bazaar moved to the opposite side of the road and it has been expanding ever since. The infrastructure of the bazaar is almost nonexistent. The water and electricity supplies are subject to constant interruption. The roads are in a very bad shape partly because they are used by heavily loaded trucks. The bazaar is open every day from 7:00 a.m. to 6:00 p.m. In 2009, due to the major highway construction, the Airport bazaar was demolished and moved to a new location with better infrastructure.

Chorsu bazaar is conveniently located in the city center of Tashkent, Uzbekistan around a large dome built during the Soviet Union period. Food and vegetables have been sold inside the dome since then. On the other hand, the surrounding area is covered with new business centers, malls, as well as tents, shelves, stalls, and containers. The bazaar continues to expand in all directions and is open every day of the week. Since the bazaar is located at the city center, transportation is very easy. The infrastructure of the bazaar is far better compared to the other two bazaars. Chorsu bazaar is also a combination of different smaller bazaars such as a textile mall, furniture mall, and gold bazaar. The goods sold in the bazaar represent a wide variety but not as diverse as in the Airport bazaar. There are wholesalers and retailers as well as some specialty shops such as one for traditional bridal costumes. Within the borders of the bazaar, there is a mosque and a church, and next to them is a large wholesale market for religious goods, which has doubled its size in the last few years.

As the name suggests, Desert bazaar is located in the desert 30 km away from the city center Ashgabat, Turkmenistan, and 5 km from the main airport. Similar to Azerbaijan, the location is close to the airport but has been moved a few times by the government to its final location, and it is expanding in all directions in the desert. Currently, the bazaar is operating on Thursdays, Saturdays, and Sundays between the hours of 5:00 a.m. and 14:00 p.m. The bazaar takes place on sand with no infrastructure except the main entrance door and the two parallel roads built to connect it to the main road. The bazaar consists of mainly parallel corridors and every corridor is concentrated on types of merchandise such as carpet, textile, and gold. Different from the other two bazaars discussed in this chapter, there is a livestock bazaar, a car bazaar, and even a spare parts bazaar. The livestock bazaar is one of the largest in the country where cows, sheep, and camels are widely traded. The spare parts bazaar offers spare parts for Soviet machinery that is not produced anymore. This bazaar is almost nonexistent in Azerbaijan and there is only a small one left in Uzbekistan. Another difference of the Desert bazaar is that there are still handmade crafts such as carpets and jewelry available, whereas in Azerbaijan and in Uzbekistan these goods are now sold in the expensive shops for tourists at the city centers. The Desert bazaar is also an attraction site for the few tourists that visit the country but the other two bazaars do not have much to offer to the foreigners. The national pride Turkmen carpets are sold in the bazaar but are subject to extreme custom regulations.

The entrepreneurs and the people who work in the three bazaars come from many different ethnic and religious backgrounds. The Soviet system did not appreciate trade and the people who were occupied with trade were considered the lower class of the society, whereas science, arts, and sports were respected. The entrepreneurs who left for Turkey with the first flights were the people who were occupied with trade beforehand. The people who later joined or are still joining the bazaars are the educated classes of the ex-Soviet Union system. These are engineers, teachers, ex-government officials as well as many with university degrees who are working as sales people. In Azerbaijan, there are also foreigners mainly from Georgia, Turkey, China, and Chechnya who are employed in the bazaar.

4.3 Relationship between the Formal Institutions and the Entrepreneurs

This chapter focuses on the legal, regulatory, national, and international financial institutions and their relationships to the entrepreneurs in the bazaars. In addition, the bazaar management and its relationship to the entrepreneurs operating in the bazaar is described with examples. For economies that are in the process of transformation from centrally planned to market based systems, formal institutions that are essential for the large scale and sustainable development of private sector businesses are either nonexistent or inadequately focused on the needs of entrepreneurs (Smallbone and Welter, 2006). Since the trust in formal institutions is low, entrepreneurs in the bazaars are trying to avoid them and replace them with more trusted informal networks.

The formal relationship that entrepreneurs have with the legal institutions concentrates in three main activities: registration processes, tax duties, and conflict resolution. The legal system in all three countries is forcing the entrepreneurs to register in order to engage in any business activity. All three governments are encouraging their enterprises to register by offering different incentives and enticements. However, challenges for the entrepreneurs start from the first day with the initial time and cost for the registration processes required, which include permits, licensing, and certifications. Usually the microenterprises and many small businesses cannot afford or are not willing to go through such procedures. Another reason to avoid registration for every size of business is the tax regulations. The tax regimes in the last decade changed many times and tax officials are generally regarded as unreliable. Although bazaars are regulated and controlled by the state officials, it is possible to avoid the initial burden to start the business and tax regulations without officially registering. However, the nonregistered enterprises and individual entrepreneurs are being inspected more often than legal entities and are always subject to financial harassment by the state authorities. Due to the problems cited, the number of registered businesses still remains very low compared to the total number of businesses. That is one of the main reasons why the bazaars are still playing an important role for the economies of Azerbaijan, Uzbekistan, and Turkmenistan. Bazaar reflects a good mixture of registered and unregistered businesses operating together. They provide a protective environment and play a critical role for the business development of enterprises.

There is a general mistrust toward the legal system in the conflict resolution process. The main reason why the entrepreneurs cannot go to the court is that most of the contracts in the bazaars are unwritten and nonbinding. In addition, since most of the businesses are unregistered and cannot conduct business by law, they are excluded by the legal system. As a result, the entrepreneurs of any size avoid seeking justice by the judicial system and try to resolve the conflict using different informal networks.

The other formal relationship that the entrepreneurs have is with the regulatory institutions. Inherited from the Soviet Union system, the state in the three countries has an extensive but mostly malfunctioning set of regulatory rules and inspections that the enterprises try to avoid by operating in the bazaars (Yessenova, 2006). The inspections are done by tax authorities, fire departments, health departments, electric departments, etc. On the other hand, if the enterprise is located at the bazaar, the management of the bazaar deals with all the necessary inspections and the inspection fees are included in the rent. One expense that enterprises cannot avoid in the bazaar is the rental fee that is paid to the bazaar management. In Azerbaijan, most of the inspection fees are included in the rental fee paid to the management. In Uzbekistan and Turkmenistan, the enterprises are still subject to various inspections in the bazaar but less so compared to registered businesses.

The relationship with the regulatory authorities is not restricted to the bazaar environment. Despite the regulations specific to the bazaar, there are other regulations that represent challenges to the entrepreneurs. As an example, the citizens of Uzbekistan and Turkmenistan have to obtain an exit visa to leave their country regardless of their destination country. To obtain the exit visa is always challenging, problems are commonplace during reentry, and the entrepreneurs can be banned from leaving the country for many years or even forever. For this reason, entrepreneurs cannot develop long-term cross-border relationships. Also, if any kind of conflict occurs with the state, the enterprises can lose everything that they have.

The financial institutions including the banking system in the three respective countries did not exist during the Soviet Union and their developments

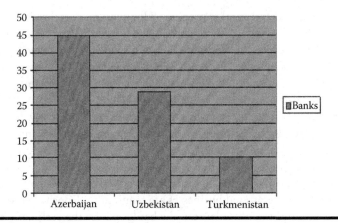

Figure 4.1   Number of banks. (adb.org, 2009, ifc.org, 2009)

showed different paces. Similar to legal and regulatory systems, there is a strong mistrust in the financial system, mainly toward the national banks (Figure 4.1).

Azerbaijan has the most sophisticated and internationally integrated financial system compared to the other two countries. There are currently 45 banks and 530 branches and the state owned International Bank of Azerbaijan dominates 47% of the market share. There are also 96 nonbank credit organizations. In Uzbekistan, there are 29 banks with 800 branches including the Central Bank. The commercial banks are divided into categories based on type of ownership: three state banks, nine private, five with foreign capital, and eleven with mixed ownership but the state directly or indirectly still has strong influence in the private banks' activities. Lastly, in Turkmenistan, there are 10 banks in total, only one foreign bank but the central bank has absolute control over all financial institutions in the country. Although by definition the central bank should act as just a regulatory body, it is the only financial entity that is allowed to have international transactions. The banking system of Turkmenistan has no contact with the outside world except through the central bank of Turkmenistan.

The formal relationship that the entrepreneurs in the bazaar have with the national financial institutions is limited. The main barriers are the registration and collaterals required by the banking system. The commercial banks can only work with the registered businesses and the unregistered enterprises are excluded by the financial system automatically. It is not possible for them to have any kind of transaction such as receiving business loans from the banks. For the registered enterprises, the banks require financial records, credit history, and collateral such as real estate, which cannot be satisfied by the majority of the entrepreneurs. In the retail banking system, the collateral is used to give credits and the most commonly accepted collateral is property. However, property ownership is still an unsolved problem for all three countries despite different legal regulations (Rona-Tas, 1994). Since the landlords cannot provide the banks with the necessary ownership documentation, they cannot receive business loans. Even when the deed exists, it is usually in someone else's name, such as a family member, to protect the business from the tax authorities. If the enterprises succeed to get into the banking system, the process is very bureaucratic and costly. It is also not easy for banks to identify the boundaries between the enterprise and the family business in the bazaar environment.

Entrepreneurs also have difficulties in access to cash, access to foreign currency, and access to credits and international money transfers. For example, in Uzbekistan, all the registered businesses are required to deposit their daily cash with the banks and the officials from the banks collect the cash daily door-to-door from every registered business. However, when the business needs the cash back from its deposit account, there is no guarantee they will receive these on the same day or even the same week. Recently, the access to cash for payroll purposes has been simplified but it remains a problem for working capital. To avoid this problem, the enterprises do not announce their daily cash correctly, which is a severe crime if detected. As a result, the trust and the relationship between the banks and Uzbek entrepreneurs are negatively affected. It is more difficult to effect such control in the bazaar but their presence is still felt by the entrepreneurs every day. To make international wire transfers for the enterprises through banks is also very difficult and they use similar unofficial ways to transfer the cash out of the country. However, when the goods arrive at customs, they have a hard time if they cannot present the official papers to declare the value of the goods. Then they have to go through more unofficial payments to receive the goods.

The international finance institutions such as International Finance Corporation (IFC), European Bank and Reconstruction and Development (EBRD), and Asian Development Bank (ADB) support the entrepreneurs with direct projects such as microfinance programs or indirect projects such as advisory services to the national banks. The relationship that the entrepreneurs have with the international financial institutions is directly linked to the cooperation between the government and those institutions (Figure 4.2).

As an example, since their independence, EBRD has completed 77 projects in Azerbaijan, 44 in Uzbekistan, and only 7 in Turkmenistan. As the number of the projects suggests, more the relationship countries have with the international financial institutions, the better their financial system development is.

Since it was first introduced in 1976 in Bangladesh by Mohammed Yunus, microfinance became a financial tool to serve the underserved entrepreneurs around the world and have served 80 million people in developing countries.

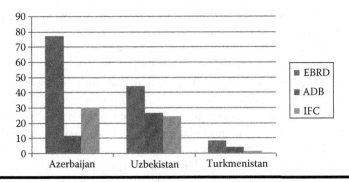

Figure 4.2 Completed and ongoing projects by IFC, EBRD, and ADB. (adb. org, 2009, ifc.org, 2009, ebrd.org, 2009)

Microfinance is proven to be an effective tool to reduce poverty, to increase employment, reduce immigration, and develop entrepreneurship. In the three focus countries, as mentioned in the previous chapters, commercial banking can only provide financial services to the registered or legal enterprises; many micro and small businesses in the bazaars were excluded from the mainstream financial services. In addition, the main stream banks are not interested in serving these entrepreneurs because of the high transaction cost and high risk associated due to not having the collateral. Thus, microfinance institutions supported by the international financial institutions took advantage of this problem by developing a new approach and offered business loans to enterprises. In this approach, they combined the financial products such as credits, loans, savings, and money transfer to serve the entrepreneurs who are not registered and cannot provide collateral. Instead of traditional collateral such as property, gold is often requested. In many cases, entrepreneurs offered gold collected from their female family members as collateral to receive business loans.

Azerbaijan, similar to its financial services, has the most advanced microfinance financial services compared to the other two countries. Microfinance projects that supported the micro, small, and medium enterprise development have shown excellent results in Azerbaijan especially during the last 5 years. However, early microfinance projects go back to mid-1990s, and in the beginning, there were many failed projects.

Microfinance projects delivered by national and international organizations have shown great rates of success working with the financially underserved enterprises. Majority of the microfinance institution clients never had prior relationships with any financial institutions and the rest had very bad experiences. The data reveals that microfinance in Azerbaijan is becoming more efficient and increasing in its outreach despite difficulties such as high inflation, absence of a specific law on microfinance, and the reluctance of local banks to provide funding (AMFA,* 2008).

One success story from Azerbaijan is the Microfinance Bank of Azerbaijan founded in 2002 by the international shareholders EBRD, IFC, the Black Sea Trade and Development Bank, the German government's development agency KfW Development Bank, and the German consulting company LFS Financial Systems GmbH. In 2004, the bank opened its third branch at the Airport bazaar. The bank was the first formal financial institution in the bazaar. In addition to the physical obstacles, such as constant electricity interruption, the bank management and its staff were challenged by the fierce competition from the informal finance institutions that were operating in the bazaar. However, the most important challenge for the management was to gain the trust of the entrepreneurs in the market. The entrepreneurs had no trust in the formal financial institutions and since most of them had no registration or collateral, they did not think it was possible for them to work with banks and receive loans. After months of direct marketing, the bank had its first clients. Today the bank has 13 branches in Azerbaijan, leads the market in microfinance with 42% of market share, and the Airport bazaar branch remains one of the biggest branches in terms of loan portfolio. In addition, the bank is the highest rated private bank in Azerbaijan by Fitch's rating in 2008. In 5 years, the bank managed not only to gain the trust of the entrepreneurs in the Airport bazaar but supported their business development. A recent decision of the Azeri government to relocate the bazaar has forced the bank to close its branch at the Airport bazaar. As this case study suggests, the bazaar traders and owners whose livelihoods depend on the bazaars are at odds with high-powered political and economic figures in informal, behind-the-scenes struggles for control of this lucrative sector of the economy (Spector, 2008b).

The microfinance industry in Uzbekistan, despite its large population and demand from the micro and small enterprises, still remains underdeveloped mainly due to the regulatory framework and operating environment. There are 32 institutions providing microfinance with a total number of 93,500 active borrowers and an outstanding portfolio of 890 million USD as of 2006.† The existing legislative framework imposes several restrictions on the microfinance institutions regarding their day to day operations in areas such as interest rates, loan size, borrower legal status, etc. On the other hand there have been many international advisory projects such as EBRD to the state and private banks to develop their credit system so they can include the micro and small business in their portfolio. Microfinance projects are still at the early development stage in Uzbekistan.

---

* *Source*: Azerbaijan Microfinance Association.
† Microfinance Information Exchange (MIX, November 2006).

Although there is need and high demand for the microfinance projects in Turkmenistan, there have been no significant international microfinance projects as of 2011. Due to the highly protective and centralized government policies, Turkmenistan has been isolated from the international finance community, including the microfinance sector. Due to more open policies of the Turkmenistan's new President, in the last 2 years there have been some efforts made to integrate the Turkmen economy with the market economy. In March 2009 United Nations Development Programme (UNDP) and other international agencies held a conference* in Ashgabat. In the conference, the lack of finance to the micro and small businesses were recognized as a main obstacle for the private sector development in Turkmenistan. The future of any microfinance program in the country totally depends on the willingness of the Turkmen president and the government to develop this sector.

Another formal relationship that entrepreneurs have in the bazaar is the relationship with the bazaar management and the bazaar owners. However, the information about the bazaar owners are often shrouded in mystery and the official administrator in the bazaar office is not the real owner, and the owner often has important political connections (Spector, 2008b). Most of the retail space in the bazaar belongs to the government and the rest to powerful groups or monopolies connected to the government. Most of the enterprises are actually renting from the government or such groups. The main responsibility of the management is to deal with rental arrangements in the bazaar. The existence of any business in the bazaar depends on the personal relationship, reputation, and unofficial payments that they make to the management. To be on the blacklist of the bazaar management is not desirable since the rental arrangements can be cancelled any time without any reason. In case of misbehavior, the relationship is damaged and it will be almost impossible to reenter the bazaar.

The following example for the Airport bazaar illustrates how much power and influence the bazaar management has over the daily business of entrepreneurs. It is estimated that there are more than 10,000 sales points located at the Airport bazaar. The proximity of the sales points to the bus stop is critical for the business and it is very important not to lose the preferred sales points in the bazaar. Closest location to the bus stop attracts more customers and represents a higher rent premium for the entrepreneurs. The bus stop location was changed a couple of years ago due to political reasons by the bazaar management and as a result many businesses went bankrupt around that area. Many partnership arrangements were affected; many customers had a difficult time to relocate the business and to make repayments. The sales points close to the new bus station was given to the entrepreneurs that were closer to the bazaar management.

---

* *Source*: www.undp.org

4.4 Relationships between the Informal Institutions and the Entrepreneurs

In an unstable environment characterized by a weak state, deficient legal regulations, and insufficient financial system, informal networks often play a key role in helping entrepreneurs to mobilize resources, create market opportunities, and cope with constraints imposed by highly bureaucratic structures (Smallbone and Welter, 2001). The institutional trust reposed in the state authorities or in the financial system is low in the three countries of interest. In addition, in an uncertain environment where there are no legal and binding contracts, the trust and the commitment to do business is highly risky for all the entrepreneurs. To reduce their risk, the entrepreneurs use certain trust networks. Social networks are often the most reliable and efficient ways of dealing with pervasive uncertainty and day-to-day business problems for entrepreneurs (Ozcan, 2006). In addition, the cost of doing business is reduced by partnering with these networks. These trust networks can be based on family, gender, ethnicity, tribe, or even religion. They are used during day-to-day operations but have become extremely important in the case of conflict resolution. As an example, almost no conflicts of any kind in the bazaar are taken to the court. Entrepreneurs feel more vulnerable in front of the state institution and they still prefer to lose their case or use alternative methods such as trust networks. For this reason, the networks develop their own set of rules to solve the conflicts. Of course the result can never be too independent of clandestine relationships and it is possible to influence the result. In time, circles of trust become very powerful and start acting as institutions and as a result monopolies are born.

The malfunctioning system and the mistrust toward the formal institutions have created many informal institutions. In the three bazaars, personal trust is much more powerful than institutional trust. The relationships that exist between the entrepreneurs and the informal institutions or networks are more difficult to identify and are more complicated to analyze than their relationship with formal institutions. Such a relationship is based almost solely on long-term trust between, reputation among, and recommendation from both parties. Such circles of trust can be analyzed in a family, gender, religion, or ethnic group and they play a critical role in conflict resolution. Another important informal relationship is the one that the entrepreneurs have with the buyers and suppliers.

The family network is extremely important especially at the initial stage for every entrepreneur operating in the bazaars and the existence of family networks is strong in all three bazaars. Many successful large enterprises in the bazaar are family businesses. In an environment with many uncertainties, working with family members reduces risk. When one family member makes enough profit to enhance the business, opens a second sales point, or decides to travel to

Turkey or China, she invites another family member to join. As an example, in the Desert bazaar, five brothers are working in five different sales points, all in leather products. The financing and transportation costs of doing business are lowered in such networks. In addition, the lack of legal contracts and conflict resolution problems increase the importance to work with family networks.

An important feature of trade during the Soviet period was the importance of female entrepreneurs. During that period, they would arrange appointments with a group of clients in their apartments to trade goods that came across borders. In addition, after the Soviet Union dissolved, many males went to Russia and other countries to work and never came back or stopped sending support for their families. As a result, a new generation of female entrepreneurs developed. There are many successful female entrepreneurs that have large businesses in all the three bazaars. In addition, it is relatively easy for the female entrepreneurs to travel across borders and deal with the custom officials. They travel abroad together and they usually bargain together to reduce their costs. As an example, in the Desert bazaar, certain sections of the bazaar such as textile and gold are dominated by female entrepreneurs. Like many other imports, gold import to Turkmenistan is highly restricted. Upon arrival at Turkmenbashi International Airport, the custom welcomes any visitors by checking if they are carrying gold with them. During the 3h short flight from Istanbul to Ashgabat, Turkmen female entrepreneurs put on as many gold necklaces, earrings, bracelets, and rings on themselves in order to claim the gold as their own so that they can avoid custom fees or even confiscation.

While the Soviet Union was in power, the role of religion in the society was minimal. However, in the recent years, the influence of religion and the number of Islamic groups increased exponentially. As a result, in the bazaar, the entrepreneurs who belong to religious networks also increased and new informal networks were formed among the members of such groups.

There are also ethnic trust networks in the bazaar. As an example, people from Dagestan have a long reputation and experience in jewelry making in the region. As an ethnic group living in Azerbaijan, in the Airport bazaar and also in Baku, their business is concentrated in gold and diamond and they operate within their closed ethnic networks. Another example for ethnic groups operating in Uzbekistan is Tajik ethnic networks. Throughout their history and also during the Soviet Union, Tajiks always dealt with trade related activities and after the break down, they became very successful entrepreneurs. Their close ethnic networks allowed them to operate more effectively in the Chorsu bazaar. There are also ethnic trust networks in the Desert bazaar such as Uzbeks and Tajiks. Since it is very difficult to get any state role for the minorities in Turkmenistan, and there are not many industries they can work in, they built their own networks in the bazaar.

Another type of informal network that exists in the bazaar is the tribal networks. Turkmenistan consists of five different tribes: Teke, Yomut, Arsary, Chowdur, and Saryk and the tribal networks are very strong. For example, the entrepreneurs from the southeast, which is close to the ancient city of Merv, are known to be traders throughout history since the city of Merv was located on the ancient Silk Road. Today they have one of the strongest trade networks in the bazaar. On the other hand, the tribe that the last two presidents of the country belonged to is the strongest tribe in the country and their networks also play an important role in the bazaar.

The informal relationship between both international and domestic suppliers and the entrepreneurs also plays an important role in the three bazaars. The domestic suppliers usually belong to the informal networks cited earlier such as ethnic or based on reputation and recommendations. On the other hand, the relationship with the international suppliers is more complicated since the contracts are also nonwritten and nonbinding. The first origin of all the goods that arrived to the three bazaars was Turkish since Turkey was the closest western country in the region. In addition, Turkey also has very close historical, cultural, language, and religious ties with all the three countries. Entrepreneurs went to Turkey mainly to Istanbul with cash or sometimes with goods such as caviar, or ex-Soviet Union products such as watches, radio, etc., and in return, they filled their suitcases with textile, shoes, etc. In time, the suitcases were replaced with containers and trucks. The Laleli district, on the historic peninsula of Istanbul, has become the primary marketplace where predominantly female shuttle traders (*chelnoki*) from the former Soviet Union meet Turkish small-scale entrepreneurs who have set up shop there to cater to the traders' demands (Yükseker, 2004). Due to the long-term relationship, Turkish suppliers and the entrepreneurs from the three bazaars developed sophisticated trading partnerships. However, in recent years, the trend changed drastically toward China. Instead of buying the cheaper Chinese goods from Turkey, entrepreneurs went directly to China. Although there are more barriers to trade in China than Turkey such as language, good business opportunities overcame these barriers. The other main reason why the trade shifted to China is better supplier credit arrangements offered by Chinese business to entrepreneurs. Entrepreneurs developed strong trading relationships with their Chinese partners and the contracts remain nonbinding and unwritten. The main destination in China for Azeri entrepreneurs is Guangzhou, Shanghai, and Hong Kong. On the other hand due to the proximity, the main Chinese destination for Uzbek entrepreneurs are the trade centers in Xinjiang-Uyghur region. Due to the highly protective policies of Turkmen Government, the international trade is relatively low compared to other two countries. However, similar to other countries, the main international trading partner of Turkmenistan is Turkey. Turkish goods and Turkish businessmen operating in the country play a critical role for the economic development of the country. In the last 5 years, Chinese goods and businesses are slowly appearing in the economic scene. This is mainly caused by an agreement to build a natural gas pipeline from Turkmenistan to China. The two governments also agreed on economic cooperation. The other important international trading relationships that the entrepreneurs developed are with United Arab Emirates, Russia, and Iran.

The relationship that exists between the entrepreneurs and their clients is also informal since even the supplier credits they provide are not based on contracts. There are wholesalers as well as retailers operating in the bazaar. The clients can be domestic as well as international. As an example, the clients of Airport bazaar come from everywhere in Azerbaijan and the bazaar is a big trade center for the neighboring countries such as Georgia, Dagestan, and Chechnya. These international relationships are often interrupted by the political instability in the region. The borders are closed or custom regulation is changed. The daily business of the entrepreneurs is directly affected by the political stability of the region.

As mentioned in the previous section the existing financial system does not meet the need of the entrepreneurs and excludes many of them and as a result, sophisticated informal financial networks developed such as the networks that transfer money across borders. To complete an international money transfer, the enterprises must be registered, the limits are very low, and the process is expensive, bureaucratic, and slow. As a result, majority of the international money transfers from the bazaars are completed through the informal networks and not through the banks. As an example, at the Airport bazaar, there are small exchange offices that are located in almost every corner and they are a part of a big network connected to the bazaar management, which plays the role of the financial institutions in the bazaar. They lend cash and charge an interest as high as 10% per month. This office also acts as a safe and keeps the cash on behalf of its clients in the evening. The excess cash is lent back to the enterprises as business loans. The transfer of the cash to Turkey or China is also conducted from these offices. When the cash is received as a payment for the goods that come from Turkey, the exchanger makes a phone call to his counterpart in Turkey. The confirmation is made over the phone, and the recipient in Turkey makes the payment to the supplier on behalf of the client in Azerbaijan and the goods are delivered. There is approximately 1% service fee for the transaction, which is subject to change depending on country and amount. The actual write-off of these transactions between parties in different countries is probably part of highly complicated organizational networks, which happens outside of the borders of the Airport bazaar. In Turkmenistan, money transfer across the borders of the country remains extremely challenging and is almost not possible for the entrepreneurs. In addition, the communication system is also not integrated with the outside world. The phone lines are not functioning well and it is believed that it is constantly monitored by the State. This situation makes it more difficult for the entrepreneurs to complete transactions even over the phone like their colleagues in Azerbaijan and Uzbekistan. In many cases the hard cash is physically carried by the entrepreneurs across borders that make them vulnerable in front of the custom officials.

Another financial problem faced by the entrepreneurs is the access to foreign currency. All the post-Soviet Union countries in the past had difficulties in managing their foreign currency flow and regulating the currency rate. Although Azerbaijan managed to solve this issue with its recent oil boom, it is a major concern for enterprises in Uzbekistan and Turkmenistan. The state regulates the amount of foreign currency available to its citizen and its rate. Majority of the legal currency exchange offices in Uzbekistan and all in Turkmenistan are controlled by the state banks. During the high trading seasons where the demand for the foreign currency is higher, there is a shortage of foreign currency. For this reason there is a "secondary or black" or "informal" foreign exchange market that provides foreign currency with different rates to the entrepreneurs. However, this market is also under constant surveillance by the state officials and their activities can be interrupted by the state at any given time.

4.5 Conclusion

The development and growth of the enterprises of any sizes in Azerbaijan, Uzbekistan, and Turkmenistan are extremely important for their economies. The policies chosen by the government have a direct impact on the business development of the enterprises. Property struggles and redistributions are an enduring feature across much of post-Soviet world; the lines between politics and business remain blurred, and political figures have significant stakes in the economic assets of these countries (Barnes, 2006). More open policies have a better impact on the business such as in the case of Azerbaijan and the opposite of such policies results in much slower development such as in the case of Turkmenistan. Nevertheless, since their independence, despite different policies, in all the three countries the number of enterprises has continuously increased.

Though the transition countries have chosen different paths of development, they have all undergone a tremendous amount of economic and social change; an important aspect of which has been the development of a new private sector (Aidis and Sauka, 2005). The legal and financial systems in all three countries are still in the development stage. Even in Azerbaijan, where the most progress has been made compared to other countries, the systems are weak and the governance related problems are still a major concern. One other reason why Azerbaijan showed better development can be explained by the presence of large multinational oil companies operating in the country. Since the multinational oil companies are under constant surveillance by their shareholders and they are highly criticized by different stakeholder groups, their presence and experience combined with the other financial institution forced the Azerbaijani government to improve their legal and financial system. In Uzbekistan, some improvements have been made to integrate the legal and financial system to international standards but the results are still not satisfactory. In Turkmenistan, the situation is so extreme that the international legal system is not recognized for the businesses operating in the country and there is no financial institution that has international operations other than the central bank of the country. As a result the number of multinational corporations operating in the country remains very small and despite the rich natural resources, the country cannot attract foreign direct investment. The entrepreneurs of all sizes including the ones operating at the Desert bazaar are negatively affected.

The presence and the number of multinational corporations and international financial institutions have had positive impact on the business environment. Microfinance projects are in high demand in all three countries. If implemented correctly and continuously, it is proven to successfully work such as in the case of Azerbaijan. However, microfinance projects have not reached the entrepreneurs in Uzbekistan and Turkmenistan yet. In the future, all three governments have to encourage microfinance projects to stimulate the growth of their enterprises.

In all three countries, the bazaar environment is a combination of different trust networks such as family, ethnic, or religious and it remains an important trade center for both registered and unregistered business. Their importance only decreases when the trust between the institutions and the enterprises are established. Although the bazaar provides protective environment for the enterprises and allows them to grow, they are still vulnerable against the state authorities and the management of the bazaar.

After their independence from the Soviet Union, all three countries developed a new class of entrepreneurs that did not exist in the past. These newly formed governments and all their institutions including legal, regulatory, and financial are also developing since their independence. These first-generation entrepreneurs learned and adapted to their environment and in most cases they integrated with the market economy faster than their government. Their struggle does not consist of competing with each other but also against many different barriers that exist in their operating environment. Some of them received external support such as microfinance but most of them are still struggling alone or within their trust networks. In addition they are also isolated by the lack of technology from the rest of the world.

The development of their economy is extremely important for the nation building process for Azerbaijan, Uzbekistan, and Turkmenistan. Both the entrepreneurs and their relations with formal and informal institutions play a crucial role in the market building process. The paper concludes that the entrepreneurs have better relationship with the formal institutions where better operating environment exists and the government has more open policies; therefore the market building process will be faster, such as in the case of Azerbaijan. On the opposite spectrum, in the countries where there are more protective and centralized policies by the government and the operating environment represents more challenges to the entrepreneurs, the relationship of the entrepreneurs with informal institutions will be stronger than the formal institutions such as in the case of Uzbekistan. However, despite their struggles and challenges, the economies of all three countries are growing and as long as there is demand for their products or services, the dynamic entrepreneurial class will continue to grow their businesses in Azerbaijan, Uzbekistan, and Turkmenistan.

References

Aidis, R. and A. Sauka. 2005. Entrepreneurship in a changing environment: Analyzing the impact of transition stages on SME development. *Challenges in Entrepreneurship and SME Research*. Inter-RENT, Turku, Finland.

Asian Development Bank, 2009. Azerbaijan, www.adb.org/Azerbaijan, 6 April 2009.

Asian Development Bank, 2009. Uzbekistan, www.adb.org/Uzbekistan, 6 April 2009.

Asian Development Bank, 2009. Turkmenistan, www.adb.org/Turkmenistan, 6 April 2009.

Azerbaijan Microfinance Analysis and Benchmarking Trends Report, 2008. A report from the Microfinance Information Exchange, Inc. (MIX) and Azerbaijan Microfinance Association (AMFA) September 2008.

Barnes, A.S. 2006. *Owning Russia: The Struggle Over Factories, Farms and Power*, Cornell University Press, Ithaca, NY.

European Bank for Reconstruction and Development, 2009. *Azerbaijan*, www.ebrd.com/pages/country/azerbaijan, 6 April 2009.

European Bank for Reconstruction and Development, 2009. *Uzbekistan*, www.ebrd.com/pages/country/uzbekistan, 6 April 2009.

European Bank for Reconstruction and Development, 2009. *Turkmenistan*, www.ebrd.com/pages/country/turkmenistan, 6 April 2009.

International Finance Corporation, 2009. Europe and Central Asia, www.ifc.org/publications/, 6 April 2009.

Micro Finance Bank of Azerbaijan (Accessbank), 2009. www.accesbank.com, 6 April 2009.

Ozcan, G.B. 2006. Djamila's journey from kolkhoz to bazaar: Female entrepreneurs in Kyrgyzstan. In *Enterprising Women in Transition Economies*, eds. F. Welter, D. Smallbone, and N. Isakova. Ashgate, Burlington, VT, pp. 93–116.

Pomfret, R. 2006. *The Central Asian Economies since the Independence.* Princeton University Press, Princeton, NJ.

Rona-Tas, A. 1994. The first shall be last? Entrepreneurship and communist cadres in the transition from socialism. *The American Journal of Sociology*, 100(1):40–69.

Smallbone, D. and F. Welter. 2001. The distinctiveness of entrepreneurship in transition economies. *Small Business Economics*, 16:249–262.

Smallbone, D. and F. Welter. 2006. Exploring the role of trust in entrepreneurial activity. *Entrepreneurship Theory and Practice*, 30(4):465–475(11).

Spector, R.A. 2008. Bazaar politics: The fate of marketplaces in Kazakhstan. *Problems of Post-Communism*, 55(6):42–53.

United Nations Development Programme, 2009. [Online], available at: www.undp.org

Uzbekistan Banking Association, 2009. www.uzibor.com, 6 April 2009.

Welter, F., T. Kautonen, A. Chepurenko, E. Malieva, and U. Venesaar. 2004. Trust environment and entrepreneurial behavior: Exploratory evidence from Russia, Germany, and Russia. *Journal of Enterprising Culture*, 12(4):327–349.

Yessenova, S. 2006. Hawkers and containers in Zarya Vostoka: How bizarre is the post-Soviet bazaar? In *Markets and Market Liberalization: Ethnographic Reflections*, eds. N. Dannhaeuser and C. Werner. Research in economic anthropology, Vol. 24, pp. 37–59, JAI Press, New York.

Yükseker, D. 2003. *Laleli-Moskova Mekigi: Kayıt dışı ticaret ve cinsiyet ilişkileri.* [The Laleli-Moscow Shuttle: Unregistered trade and gender relations] Iletişim Yayınları, Istanbul, Turkey.

Yükseker, D. 2004. Trust and gender in a transnational market: The public culture of Laleli, Istanbul. *Public Culture*, 16(1):47–65.

Chapter 5

Israel: A Start-Up Nation in a Global Supply Chain Context: The Revival of a Virtual Silk Road

Ehud Menipaz

Contents

5.1　Introduction

The State of Israel, located at the Eastern Mediterranean Basin, was established in 1948, after nearly 2000 years of Jewish dispersal. The 63 years since the State of Israel was created have been marked by an accelerated infrastructural, economic, social, and technological development. This development has been made despite on-going conflict with neighboring Arab states, culminating in several wars and leading to peace agreements with both Egypt and Jordan. It seems that the remarkable development of the Israeli economy was made possible partially because of a revival of a virtual Silk Road. By developing appropriate economic, business, transportational, technological, educational, political, and social ecosystem, Israel has defined itself as a relevant link in a global supply chain where research and development (R&D), entrepreneurship, and innovation are recognized, encouraged, and supported.

Historically, the Eastern Mediterranean Basin has been a very favorable international logistics hub. Many ancient routes were charted here such as the Silk Road, the Kings Way, and the Spice Road. For thousands of years these ancient routes facilitated East–West trade, cultural exchange, and religious influence of historical significance. In the land of Israel numerous landmarks, both natural and man made, marked these routes. In time, the ancient roads gave way to more modern forms of transportation by sea, air, and land.

It is argued here that old trading routes have not disappeared. They simply transformed to become global, sometimes virtual, value chains. These value chains are developed and nurtured by Multinational Enterprises (MNE) on the one hand, and developed, nurtured, and supported by national, regional, and local governments, on the other*. Israel, through its innovation, entrepreneurship skills, economic orientation, and government resolution has managed to become relevant in the context of global supply chains, be it in telecommunication, medical devices, irrigation technologies, or security services. As such, Israel serves as a hub for R&D, entrepreneurship, and innovation. The R&D,

---

* For the evolution of MNE's global supply chains, one may consult Menipaz, E., Menipaz, A., 2011. *International Business: Theory and Practice*, London, U.K. Sage Publications, Chapters 1, 9, 11, 14.

entrepreneurship, and innovation focus result in new and innovative products, new production processes, and innovative services, feeding manufacturing facilities in China and the Far East, and using global distribution channels to target markets in Europe, North America, South America, and the Far East. The Israeli case underlines the fact that vibrant national ecosystems cannot be sustained by operating in isolation and should rely on modern versions of the Silk Road that provide the synergies and the infrastructure required for a global supply chain.

The following paragraphs provide an overview of Israel as a link on a global value chain, including a brief description of Israel's economic, business, logistics, and technology ecosystems. Particular focus is placed on the country's entrepreneurship and innovation phenomena in a global context and a discussion on the high technology industry, which constitutes a significant part of Israel's economy and exports.

STATE OF ISRAEL: ESSENTIAL STATISTICS*

Population: 7.3 million

Land area: 8,367 square miles (21,671 square km)

Arid/desert land area: 4,633 square miles (12,000 square km) (more than half the country's land area)

GDP (PPP): $201.3 billion

GDP growth: 4.9% (5 year compound annual growth)

Unemployment: 6.2%

Inflation (CPI): 3.90%

FDI inflow: $9.6 billion

Population: 7.23 million

Ease of doing business index: 29 out of 183 (2010)[†]

Global competitiveness index: 24 out of 133 countries (2010)[‡]

Economic freedom index: 44 out of 183 countries (2010)[§]

---

[*] Central Bureau of Statistics, Government of Israel, http://www1.cbs.gov.il/reader/cw_usr_view_Folder?ID=141

[†] The World Bank, 2010. http://data.worldbank.org/indicator/IC.BUS.EASE.XQ

[‡] *Global Competitiveness Report 2010_2011.* http://www3.weforum.org/docs/WEF_GlobalCompetitivenessReport_2010-11.pdf

[§] The Heritage Foundation, *2011 Ease of Doing Business*, http://www.economicfreedom/index/

OECD membership: since 2010

Capital city: Jerusalem (+2 GMT)

Currency: New Israeli Shekel (ILS)

Languages: Hebrew, Arabic, English (most commonly used foreign language)

Religions: Jewish 76.4%, Muslim 16%, Christians 2.1%, Druze 1.6%, other 3.9%

5.2 History

Since gaining independence, on May 14, 1948, the State of Israel has maintained an accelerated infrastructure, economic, social, and technological development. The State of Israel is located at the Eastern Mediterranean Basin, which was always a very favorable international logistics hub. Many ancient routes were charted throughout the land, marked by numerous landmarks, both natural and man made. For example, in the north of the country, mount Tabor is believed by some scholars to be where the site of the Transfiguration of Jesus is located. The Tabor mount was part of the ancient Silk Road. Even before the official establishment of the Silk Road in the second century BC, silk was already a prized commodity in this area and elsewhere. Silk production was a closely guarded secret—which the Chinese kept to themselves for a very long time, leading to the silk trade with the West. It is noted that in modern times, during the late nineteenth century, Baron Edmond de Rothschild tried unsuccessfully to introduce silk making close to the Tabor mount and the upper Galilee area. The ancient King's Way is located in the south of the country. The King's Way was a trade route of vital importance to the ancient Middle East traders and armies.* It began in Heliopolis, Egypt, and stretched across the Sinai Peninsula to Aqaba, in the current Hashemite Kingdom of Jordan.† From there it turned northward, leading to Damascus and the Euphrates River and joining the Silk Road to China.‡

Numerous ancient nations, including Edom, Moab, Ammon, depended largely on the King's Way for trade.§ Many of the wars during the period of the Kingdom of Israel (and its sister-kingdom, the Kingdom of Judah) were over

---

* http://www.shortopedia.com/A/N/Ancient_Near_East
† Ibid.
‡ Ibid.
§ http://answers.yahoo.com/question/index?qid=20090119123515AAsmBeb

control of the King's Way. The Nabateans used this route as a trade route for luxury goods such as spices from southern Arabia. During the Roman period, the King's Way was rebuilt and called the *Via Traiana Nova*. The King's Way has also been used as an important pilgrimage route for Christians, as it passed many sites important in Christianity, including Mount Nebo, where Moses was believed to be sighted last, and the Baptism site at the Jordan River, where Jesus is believed to have been baptized by John the Baptist.* Muslims used the King's Way as the main Hajj route to Mecca.

5.3 Silk Road and International Trade

As noted earlier, the local routes in the country were feeding the historical Silk Road around the first century, linking the Eastern and the Western worlds. The Silk Road, often under the presence of hostile tribes, induced the creation of various payment and letter of credit schemes. It has been said that as convoys destined for China left Spain, commitments to pay upon receipt of merchandise by the buyers in Spain were drafted by Jewish religious leaders (Rabbis) addressed to their counterparts in China, be it in Shanghai, on the eastern seaboard of China, or the ancient city of Kaifeng in central China (Xu, 2003). For security, camels in convoy were fed with canisters full of gold coins for payments, in order to avoid being robbed on the long way to and from China. The Spice Road was another trade road that passed through the southern part of the current State of Israel, in the desert area, called the Negev. In the fifteenth century, trade along the land-based Spice Road was replaced by marine routes discovered by Vasco da Gama in 1498. At that time, the spice trade between East and West replaced silk trade as the main trading good. Modern times saw development of newer means of transport and free trade agreements, which altered the logistics approach prevalent during the Middle Ages. Newer means of transport led to the establishment of new routes, and countries opened up borders to allow trade. Nowadays the international trade is monitored and managed through bilateral and multilateral arrangements, including World Trade Organization covenants. It is argued here that old trading routes were reopened during modern times, as nations have become links on a global, sometime virtual, value chains. This fact gave way to two major characteristics of international business: First, new global value chains were developed by MNE. Second, countries do their best to be attractive to regional headquarters of MNEs as explored by Finger and Menipaz (2008) and Lowengart and Menipaz (2002).

---

* Ibid.

5.4 Economic and Business Ecosystems

Israel is a parliamentary democracy and practices free market economy along with well-developed social state policies, including state sponsored health and retirement schemes.

Despite the world economic crisis of 2008–2010, the local economy has expanded its Gross National Product (GNP) during the past 5 years and has diversified its productive base. According to the Index of Economic Freedom (Miller and Holmes, 2011), Israel is the 44th freest in the world, out of 183 countries. In the subindices of fiscal freedom and investment freedom, as defined by Miller and Holmes (2011), Israel is ranked 5 out of 17 in the Middle East North Africa (MENA) region, making Israel a likely candidate for entrepreneurship, venturing, and innovation.

Israel is open to global trade, namely export and import activities, as well as to portfolio investments, through the capital market, and to foreign direct investment (FDI), through direct purchase of non financial assets. While foreign investment is encouraged, it is restricted in a few sectors, such as defense. Regulations on acquisitions, mergers, and takeovers apply equally to foreign and domestic investors.

As for international trade, the cost of trade is burdened with high agricultural tariffs, importation fees and taxes, a complex and not transparent tariff rate quota system, strict labeling rules, import licensing and taxes, and nontransparent government procurement.[*] Israel supports its overall economic competitiveness by robust protection of property rights and relatively low corruption rate.[†] As described by Miller and Holmes (2011), Israel is committed to economic restructuring and development, replacing traditional industries with export oriented innovative and high technology ventures. Technology companies, such as Amdocs, which specializes in customer experience systems, are making global inroads with high technology products. However, companies such as Teva, the largest generic drugs manufacturer in the world, make their global mark with either innovative production processes or highly effective registering, licensing, and distribution systems.

Government spending is about half of GDP and income and corporate tax rates are relatively high. The top income tax rate is 46% and the maximum corporate rate is 26%.[‡]

---

[*] For a complete discussion refer to http://www.heritage.org/index/Country/Israel

[†] http://www.heritage.org/index/Country/Israel

[‡] For a complete discussion refer to http://www.scribd.com/doc/25509373/Index-2010-Economic-Freedom-Chapter7

Other taxes include a value added tax (VAT), currently at 16% and also a capital gain tax. A very important part of an appropriate national business eco-system is the bureaucracies involved in the creation and sustainability of busi-nesses. Starting a business in Israel takes an average of 34 days, slightly less than the world average and obtaining a business license requires slightly more than the world average of 18 procedures and 218 days.* Venturing may take the form of acquiring state owned assets and companies and renewing them. There were ample opportunities in manufacturing, service, financial industry, and telecom-munication and transportation industries. Privatization has been accelerating in recent years in all areas, including the state owned national airline, El Al and the national telecommunication corporation, Bezeq. Venturing applies to the energy sector as well. While the energy sector is state owned and highly regulated, there have been numerous drilling gas and oil ventures by private entrepreneurs. The latest discovery of significant reserves of natural gas in Israel's "Economic waters" in the Mediterranean Sea led the way to an arbitrary and ret-roactive corporate tax hike, which may deter new high risk ventures of this kind. Israel maintains conservative fiscal and monetary policies that are carefully pre-sented to and monitored by global economic forums, such as the World Bank.

The national fiscal policy assures investors of a steady and predictable busi-ness environment.

Public debt was reduced from 100% of GDP in 2003 to below 80% in 2010. Inflation was moderate, averaging 3.5% between the years 2006 and 2009. The government controls prices through the public sector and provides subsidies par-ticularly for agricultural production.† Money supply is essential in an economy that supports entrepreneurship and innovation. The country has a developed and nurtured banking system, capital market, and venture capital funds. The bank-ing sector is highly concentrated, and the five principal banking groups together hold more than 95% of total assets.‡ Commercial banks provide a full range of financial services that facilitate intense Total Entrepreneurial Activity (TEA) and robust private-sector development. Credit is available on market terms, and financial institutions offer a variety of financial instruments. Capital markets have been largely liberalized as part of Israel's effort to reinvent itself as a finance hub.§ In light of the global economic crisis it is interesting to note the relatively lower impact of the crisis on the country's financial system. The financial market demonstrated a sharp recovery during the first half of 2009, offsetting most of the losses triggered by the global financial crisis of 2008.

---

* Ibid.
† Ibid.
‡ Ibid.
§ Ibid.

A well-developed legal system is essential for an ecosystem that supports entrepreneurship and innovation. Indeed, the legal system in Israel is modern and fashioned after the British common law. Property rights and contracts are enforced effectively. Courts are independent, and commercial law is clear and consistently applied. Commercial law is consistent and standardized, and international arbitration is binding in dispute settlements with the state. Residents and nonresidents may hold foreign exchange accounts, and there are no controls or restrictions on current transfers, repatriation of profits by MNEs, or other transactions. Labor regulations are relatively flexible. The non-salary cost of employing a worker is low, but dismissing an employee is relatively costly. In order to encourage both local and foreign investors it is essential to contain various forms of corruption. It is noted that in 2010 Israel ranks 30th out of 179 countries by Transparency International's Corruption Perceptions Index (Transparency International, 2010). Bribery and other forms of corruption are illegal. Israel became a signatory to the OECD Bribery convention in November 2008. Several nongovernmental organizations focus on public-sector ethics. To sum it up, Israel, a member of the Organization of Economically Developed Countries (OECD) since 2010, has developed economic and business ecosystems that support an entrepreneurial and innovative environment including form of government, international trade practices, business practices, legal system, and financial system. This makes the country a robust link in a global supply chain.

5.5 Transportation and Logistics Ecosystems

A necessary prerequisite to being a viable link on a global value chain is a well-developed transportation and logistics systems. The transportation and logistics ecosystems in Israel seem to be affected by both the small country's land area, which makes for short travel times and the fact that land-based transportation with adjacent countries is very limited. These two factors make for very intense local land-based transportation on the one hand and a very well-developed air- and marine-based logistics on the other. A variety of modes of transportation and transportation infrastructure are well developed, and are continuously being upgraded to meet the demands of economic expansion, population growth, security needs, and tourism. Israel's road network spans 18,096 km of roads, of which 230 km are classified as freeways, including a toll road going north–south.* Buses are the country's main form of public transportation and the bus service has been franchised to privately held companies, increasing competition

---

* http://wpedia.goo.ne.jp/enwiki/Transportation_in_Israel

and improving service. The largest central bus terminal in the country is the Tel Aviv Central Bus Station. Taxicab service is well developed and includes both regular taxicab service and share taxi service, run by several private companies. Many of Israel's railway lines were constructed before the founding of the state during the Ottoman and British rule. The first line, started in the 1880s was the Jaffa–Jerusalem railway, followed by the Jezreel Valley railway, which formed part of the greater Hejaz railway. World War I prompted the building of multiple new lines out of military needs. Portions of what is now the Coastal railway were built simultaneously by the Turkish and British and later merged during the British Mandate.* Southern lines were built from the north by the Ottomans and from the Sinai Desert in the west by the British.† As of now, major expansion of railway infrastructure and service is underway, including the construction of new stations in Tel Aviv and elsewhere, the high-speed railway to Jerusalem, an extension of the coastal railway from Tel Aviv toward the southern coastal cities and Beer Sheva, located in the southern, arid, part of Israel, the Negev region. Railway lines total 949 km. A light rail system has recently been launched in the city of Jerusalem. As of 2010, there are 48 airports in Israel, the largest is Ben Gurion International Airport located near Tel Aviv, which is the destination of most international flights to Israel. It serves more than 10 million passengers annually. Israel's largest airline is El Al Israel Airlines, a privatized government airline. Israel's civil aircraft fleet consists of more than 50 aircraft, including passenger and freighters planes, mostly Boeing jets and some Airbus, ATR, and Dash makes. There are 30 airports with paved runways, 18 airports with unpaved runways, and 3 heliports. As for marine transportation there are 7 ports and marinas in Israel and 18 ships, most of which are container ships. Many other ships are owned and operated by Israeli companies but not accounted for since they are operated under foreign flags of convenience. To wit, Israel's Zim Navigation is not only the largest cargo shipping company in Israel but also one of the largest such companies worldwide.‡ Special attention is given these days in Israel to the development of "green" transportation modes to enhance environmental sustainability, including the use of public transportation, cycling, and walking. Also, a major new initiative to introduce electrical cars on a national basis is being promoted by an organization called "A Better Place." It features Chinese cars equipped with rechargeable and replaceable batteries, which are warehoused in various locations around the country.§ In

---

* Ibid.
† Ibid.
‡ Ibid.
§ For a complete discussion of the electrical cars venture in Israel consult http://www.betterplc.co.il/

summary, the transportation and logistics ecosystems are well developed and entail a very intense local land-based transportation mode as well as intense international air and marine-based logistics systems.

5.6 Entrepreneurship and Innovation

As argued earlier, it seems that Israel has established itself as part of a renewed Silk Road, overcoming the shortage of natural resources and lack of land transportation links with neighboring countries. To a large extent, Israel's economy is based largely on technology, intensive entrepreneurship, and innovation activities that are unique and sustainable. Thus, instead of producing and trading silk, spices, citrus fruits, and other merchandised goods of generations past, it trades in high technology know-how, products, and services with countries along the Silk Road and elsewhere. One may examine the entrepreneurial ecosystem in Israel through the Global Entrepreneurship Monitor (GEM) paradigm, described by Bosma and Levie (2009) and Menipaz et al. (2010). The GEM project, based on GEM paradigm, is a longitudinal, worldwide research project of entrepreneurship conducted concurrently in more than 50 countries.* The GEM methodology includes a scientific definition of entrepreneurship, an econometric model of entrepreneurship, and an extensive national and international data collection, including both random sampling and in depth interviews. The entrepreneurship phenomenon is characterized and national policy implications are structured. Among the main findings are measurements of early-stage TEA, inclusive of nascent entrepreneurs and new business owners, entrepreneurship financing, social entrepreneurship, entrepreneurship education and training, entrepreneurship among immigrants and minorities, female versus male entrepreneurship, and intrapreneurship, within company entrepreneurship. The observations further discriminate between necessity-based entrepreneurship versus opportunity-based entrepreneurship, indicate how much of this economic activity is export oriented, assess the job creation potential of the same, relate entrepreneurship to the rate of innovation, provide insight into academic education and training, and estimate the effect of government bureaucracy on national entrepreneurial prevalence. The following sections describe the methodology behind GEM and the three dynamic interactive components of entrepreneurship (entrepreneurial attitudes, activity, aspirations) and provide a comparative analysis of entrepreneurship in Israel relative to other countries.

---

* For details consult www.gemconsortium.org

5.7 GEM Model*

Business entrepreneurs drive and shape innovation, speed up structural changes in the economy, and introduce new competition, thereby contributing to productivity. The GEM research paradigm includes a distinction among phases of economic development, in line with Porter's typology of "factor-driven economies," "efficiency-driven economies," and "innovation-driven economies" (Porter et al., 2002). Necessity-driven self-employment activity tends to be higher in less-developed economies. Such economies are unable to provide demand-sufficient jobs in high-productivity sectors, and so many people must create their own economic activity. As an economy develops, the level of necessity-driven TEA gradually declines as productive sectors grow and supply more employment opportunities. At the same time, opportunity-driven TEA tends to pick up with improvements in wealth and infrastructure, introducing a qualitative change in overall TEA and self actualization sentiments. Since entrepreneurial activities vary with economic development, national policy makers need to tailor their socioeconomic programs to the development context of their country. In factor-driven economies the focus of government actions is on providing basic requirements needed to enhance entrepreneurship. In efficiency-driven economies the focus of public policy is on efficiency enhancers. In innovation-driven economies the basic requirements should exist, but the focus of public policy is on entrepreneurial conditions. As an innovation-driven economy, Israel has been nurturing entrepreneurial conditions, for the last two decades, through government-sponsored "technology greenhouses," government-sponsored venture capital, tax incentives, and the creation of a small and medium size business authority, among others. Several statutes and laws support the creation and sustainability of small and medium size businesses.

5.7.1 Entrepreneurship: Attitudes, Activity, and Aspirations†

The GEM model recognizes that a range of environmental conditions affect three main components of entrepreneurship: attitudes, activity, and aspirations, which in turn fuel the new economic activity, generating jobs and wealth. *Entrepreneurial attitudes* are attitudes toward entrepreneurship. For example, the extent to which people think there are good opportunities for starting a business,

---

* This section is based on GEM findings. For details see http://www.scribd.com/doc/26821533/Global-Entrepreneurship-Monitor-2009-Global-Report
† http://www.scribd.com/doc/26821533/Global-Entrepreneurship-Monitor-2009-Global-Report

or the degree to which they attach high status to entrepreneurs (Yedidson and Menipaz, 2010). Israelis hold entrepreneurs in high regard, almost as cultural heroes and the Israeli media provides continuous coverage of successful business exits. *TEA* is the extent to which people in a population are creating new business activity, both in absolute terms and relative to other economic activities. It is very common for many Israeli citizens to engage in early entrepreneurship activity; many do it as an extension of their day job or as a favorite social activity. *Entrepreneurial aspiration* reflects the entrepreneurs' aspirations to introduce new products and production processes, to engage with foreign markets, to develop a significant, sustainable organization, and to fund growth with external capital. Israeli citizens aspire to become part owners of a fast growing business, deploying intellectual property to introduce new products and services worldwide. As a matter of fact, growth expectations of many Israeli entrepreneurs were found through GEM to be significantly higher than those in many other countries. The GEM methodology provides measures for all of the aforementioned dimensions of entrepreneurship.

5.7.2 Entrepreneurial Framework Conditions

Entrepreneurial Framework Conditions (EFCs) reflect major features of a country's socioeconomic milieu that are expected to have a significant impact on the entrepreneurial sector. The GEM model maintains that, at the national level, different framework conditions apply to established business (EB) activity and to new business activity. The relevant national conditions for factor-driven economic activity and efficiency-driven economic activity are adopted from Schwab (2011). The GEM model demonstrates the link between entrepreneurship and economic development. As economies progress and scale economies become more and more relevant, other conditions, which are called efficiency enhancers, ensure a proper functioning of the market. GEM focuses on the role played by individuals in the entrepreneurial process and observes the actions of entrepreneurs who are at different stages of the process of creating and sustaining a business. The first phase is up to 3 months, where wages are paid, and it is termed as *nascent entrepreneurs'* phase. The second phase, between 3 and 42 months, is termed as *new business owners'* phase. The third phase, over 42 months, is termed as *EB owners' phase.*

5.8 Israeli Entrepreneurship in a Global Context

Let us examine the entrepreneurship and innovation ecosystem in Israel through the assessment of entrepreneurial attitudes, activity, and aspirations in 54

Table 5.1   Country Groups for the 54 GEM 2009 Countries

Factor-driven economies
Algeria,[a] Guatemala,[a] Jamaica,[a] Lebanon,[a] Morocco,[a] Saudi Arabia,[a] Syria,[a] Tonga, Uganda, Venezuela,[a] West Bank and Gaza Strip, Yemen
Efficiency-driven economies
Argentina, Bosnia and Herzegovina, Brazil, Chile,[a] China, Colombia, Croatia,[a] Dominican Republic, Ecuador, Hungary,[a] Iran, Jordan, Latvia,[a] Malaysia, Panama, Peru, Romania,[a] Russia,[a] Serbia, South Africa, Tunisia, Uruguay[a]
Innovation-driven economies
Belgium, Denmark, Finland, France, Germany, Greece, Hong Kong, Iceland, Israel, Italy, Japan, Republic of Korea, Netherlands, Norway, Slovenia, Spain, Switzerland, United Kingdom, United Arab Emirates, United States

[a] Country in transition to next stage.

countries, as per GEM.* The countries included in this assessment are listed in Table 5.1. The countries are grouped into three phases of economic development as defined in the Global Competitiveness Report 2009–2010 (Schwab, 2009). Israel is included in the innovation-driven economies group.

5.8.1 Entrepreneurial Attitudes and Perceptions

For TEA to occur in a country, both opportunities for entrepreneurship and entrepreneurial capabilities need to be present. However, equally important is that individuals *perceive* opportunities for starting a business in the area they live and that they *perceive* they possess the capabilities to start a business. The quantity and quality of perceived opportunities and capabilities may be enhanced by national conditions such as economic growth, population growth, culture, and national entrepreneurship policy. Table 5.2 lists several GEM indicators concerning individuals' own perceptions toward entrepreneurship for each of the 54 GEM 2009 nations.[†]

A close examination of entrepreneurial perceptions and attitudes for Israel relative to other countries may explain the prevalence of entrepreneurship economic phenomenon in the country. Perceived entrepreneurial opportunities,

---

* Ibid.
[†] Ibid.

Table 5.2 Entrepreneurial Attitudes and Perceptions in the 54 GEM Countries in 2009, by Phase of Economic Development, GEM 2009

	Perceived Opportunities	Perceived Capabilities	Fear of Failure[a]	Entrepreneurial Intentions[b]	Entrepreneurship as a Good Career Choice	High Status of Successful Entrepreneurs	Media Attention for Entrepreneurship
Factor-driven economies							
Algeria	48	52	31	22	57	58	39
Guatemala	57	64	24	18	77	69	68
Jamaica	42	77	24	29	76	77	74
Lebanon	54	77	21	22	85	79	65
Morocco	52	80	24	39	83	80	63
Saudi Arabia	69	73	49	34	80	89	78
Syria	54	62	18	54	89	89	55
Tonga	56	53	65	6	91	52	80

Uganda	74	85	29	58	81	85	74
Venezuela	48	59	26	29	76	69	49
West Bank and Gaza Strip	50	56	36	24	88	78	52
Yemen	14	64	65	9	95	97	96
Average (unweighted)	52	67	34	29	82	77	66
Efficiency-driven economies							
Argentina	44	65	37	14	68	76	80
Bosnia and Herzegovina	35	57	32	17	73	57	51
Brazil	47	53	31	21	81	80	77
Chile	52	66	23	35	87	70	47
China	25	35	32	23	66	77	79
Colombia	50	64	29	57	90	74	82
Croatia	37	59	35	8	68	49	53

(continued)

Table 5.2 (continued) Entrepreneurial Attitudes and Perceptions in the 54 GEM Countries in 2009, by Phase of Economic Development, GEM 2009

	Perceived Opportunities	Perceived Capabilities	Fear of Failure[a]	Entrepreneurial Intentions[b]	Entrepreneurship as a Good Career Choice	High Status of Successful Entrepreneurs	Media Attention for Entrepreneurship
Efficiency-driven economies							
Dominican Republic	50	78	27	25	92	88	61
Ecuador	44	73	35	31	78	73	55
Hungary	3	41	33	13	42	72	32
Iran	31	58	32	22	56	78	61
Jordan	44	57	39	25	81	84	70
Latvia	18	50	40	10	59	66	51
Malaysia	45	34	65	5	59	71	80
Panama	45	62	26	11	74	67	50
Peru	61	74	32	32	88	75	85
Romania	14	27	53	6	58	67	47

Russia	17	24	52	2	60	63	42
Serbia	29	72	28	22	69	56	56
South Africa	35	35	31	11	64	64	64
Tunisia	15	40	34	54	87	94	70
Uruguay	46	68	29	21	65	72	62
Average (unweighted)	36	54	35	21	71	71	62
Innovation-driven economies							
Belgium	15	37	28	5	46	49	33
Denmark	34	35	37	3	47	75	25
Finland	40	35	26	4	45	88	68
France	24	27	47	16	65	70	50
Germany	22	40	37	5	54	75	50
Greece	26	58	45	15	66	68	32

(continued)

Table 5.2 (continued) Entrepreneurial Attitudes and Perceptions in the 54 GEM Countries in 2009, by Phase of Economic Development, GEM 2009

	Perceived Opportunities	Perceived Capabilities	Fear of Failure[a]	Entrepreneurial Intentions[b]	Entrepreneurship as a Good Career Choice	High Status of Successful Entrepreneurs	Media Attention for Entrepreneurship
Innovation-driven economies							
Hong Kong	14	19	37	7	45	55	66
Iceland	44	50	36	15	51	62	72
Israel	**29**	**38**	**37**	**14**	**61**	**73**	**50**
Italy	25	41	39	4	72	69	44
Japan	8	14	50	3	28	50	61
Korea	13	53	23	11	65	65	53
Netherlands	36	47	29	5	84	67	64
Norway	49	44	25	8	63	69	67
Slovenia	29	52	30	10	56	78	57
Spain	16	48	45	4	63	55	37
Switzerland	35	49	29	7	66	84	57
U.K.	24	47	32	4	48	73	44

United Arab Emirates	45	68	26	36	70	75	69
United States	28	56	27	7	66	75	67
Average (unweighted)	28	43	34	9	58	69	53

Source: From Bosma, N. and Levie, J., *Global Entrepreneurship Monitor 2009 Executive Report,* London Business School, London, U.K., 2009.

[a] Denominator: 18–64 population perceiving good opportunities to start a business.
[b] Denominator: 18–64 population that is not involved in entrepreneurial activity.

entrepreneurial intentions, entrepreneurship as a good career choice, and the status of entrepreneurs are all rated higher in Israel than the average for the innovation-driven countries.

5.8.2 Entrepreneurial Activity

Figure 5.1 presents the early-stage TEA rates for each GEM 2009 country. In 2009, the entrepreneurial activity rate (ETA) for Israel was 6.0%, about double the rate in Japan, the lowest in the innovation-driven economies.

As of 2009 the TEA rate in Israel was 6.0% in 2009, compared to 6.6% in 2008 and 5.4% in 2007, possibly reflecting the effect of the global economic crisis of 2008–2009 (Menipaz et al., 2010). TEA female rate was 4.1% versus Male rate of 8.0% (see Figure 5.2). The rate of TEA among the segments of Israeli population (age 18–64) was: 6.0% among veteran Jewish people, 4.0% Russian Jewish immigrants, and 6.0% among Israeli Arabs (6.8%, 5.75%, and 6.6%, respectively, in 2008). EB ownership decreased slightly to 4.17% (4.5% in 2008). 5.9% of EB owners were male and only 1.89% females (6.2% and 2.5%, respectively, in 2008). Some 73% of entrepreneurs cite opportunity rather than necessity as their motive for creating a new venture. The year 2009 was the 1st year that Israeli women outpaced Israeli men in opportunity-driven entrepreneurship. The dispersion of TEA country estimates in Figure 5.2 demonstrates that entrepreneurship rates are not just a function of differences in

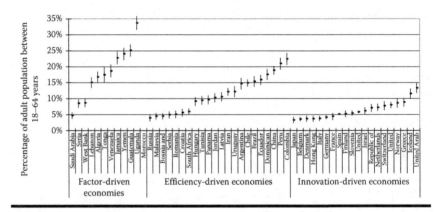

Figure 5.1 Early-stage TEA for 54 nations in 2009, by phase of economic development, showing 95% confidence intervals. (From Bosma, N. and Levie, J., *Global Entrepreneurship Monitor 2009 Executive Report*, London Business School, London, U.K., 2009.)

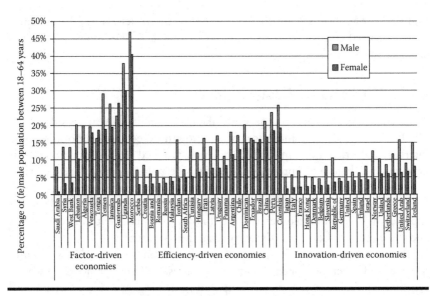

Figure 5.2 Early-stage ETAs by gender, 2009. (From Bosma, N. and Levie, J., *Global Entrepreneurship Monitor 2009 Executive Report,* **London Business School, London, U.K., 2009.)**

economic development phase but also other factors.* The GEM research project further relates country's Gross Domestic Product (GDP) to early-stage TEA level adjusted for purchasing power. It seems that many of the countries with lower GDP demonstrate a high rate of necessity-driven entrepreneurship while the countries with higher GDP demonstrate a high rate of opportunity-driven entrepreneurship. As for age group, it is reported that the 25–34 age group has the highest participation rate in every phase of economic development.† Figure 5.2 displays the differences in female and male participation for each country in GEM 2009, ordered by the major phase of economic development and female participation rate. The ratio of female to male participation varies considerably in each phase, reflecting different culture and customs regarding female participation in economic activity. Israel shows a ratio of about two male to one female entrepreneur.

Israel's institutional characteristics, demography, entrepreneurial culture, and stage of economic development seem to align with the results of GEM research

---

* http://www.docstoc.com/docs/11239765/Global-Entrepreneurship-Monitor—2008-Executive-Report

† http://www.scribd.com/doc/26821533/Global-Entrepreneurship-Monitor-2009-Global-Report

findings. GEM results confirm that institutional characteristics, demography, entrepreneurial culture, and the degree of economic welfare shape a country's entrepreneurial landscape.

5.8.3 Entrepreneurial Aspirations

As for entrepreneurial aspirations, GEM data collected in a 5 year period (2004–2009) is used to develop indicators of job-growth expectation, innovation, and international orientation in GEM countries, grouped by phases of economic development. Figure 5.3 presents the percentage of high-growth entrepreneurs who are expected to contribute in a significant way to the respective economies. Israel seemed to rate relatively higher regarding this indicator as well as the expected number of employees and the innovation component in new ventures and the intensity of export activity based on new ventures.

It should be noted that a significant amount of innovation and new ventures in Israel and other developed economies are a result of entrepreneurship within organizations, otherwise termed "corporate venturing" or "intrapreneurship." An intrapreneur is an employee developing new business activities for his or her employer, including the design and launch of new products or product/market

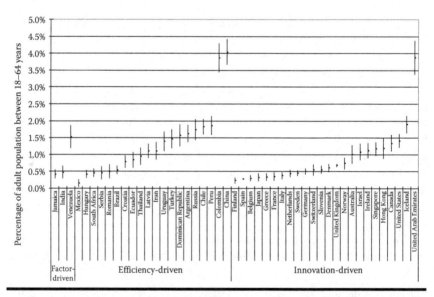

Figure 5.3 High growth expectation early-stage entrepreneurship (HEA), 2004–2009. (From Bosma, N. and Levie, J., *Global Entrepreneurship Monitor 2009 Executive Report*, London Business School, London, U.K., 2009.)

combinations. It appears that entrepreneurial activities by employees may be more prevalent in more advanced phases of economic development. Furthermore, intrapreneurship and new independent firm activity may be substitutes rather than positively correlated. That is, given a supply of entrepreneurial talent, whether individuals exploit entrepreneurial opportunities within a business or choose to start up for themselves may depend on various factors, such as the level of economic development, the institutional framework, and management styles within organizations (Yedidsion and Menipaz, 2010).

5.8.4 Entrepreneurial Framework Conditions

GEM research project assesses EFCs, which reflect the major features of a country that are expected to have a significant impact on the entrepreneurial sector. The relevant national conditions for factor-driven economic activity and efficiency-driven economic activity are adopted from Schwab (2009). The GEM model makes a contribution to the global competitiveness methodology on economic development by identifying framework conditions that are specific to innovation and entrepreneurship. Nine different EFCs are described in Table 5.3.

As per GEM research methodology, for each of these EFCs, Likert scale items were completed by at least 36 experts in each country. The results for each country are presented in Table 5.4. In general, experts in more economically developed countries assigned higher ratings to EFCs. This is consistent with the GEM model and the notion that EFCs have higher priorities among more economically developed countries. An interesting derivative of the same is that better EFCs in a country lead for a higher rate of intrapreneurs selecting that country for locating a new venture or a regional office (Lowengart and Menipaz, 2002). For Israel, the three most positive EFCs identified are: The presence of commercial and professional infrastructure, including physical, transportation and communication infrastructure and cultural and social norms.

5.8.5 Entrepreneurial Activity and the Global Economic Crisis

The global economic crisis of 2008–2010 has had an impact on entrepreneurship worldwide. However, while it has affected the amount of venture capital that was made available for new ventures, it has served to increase, at times, the level of entrepreneurship, out of necessity or because new opportunities have become available as exiting businesses were restructuring. As per Menipaz et al. (2010), during the global economic crisis Israel's unemployment rate rose from 6.2% in 2008 to 8.0% in 2009. Israel's GDP decreased in 2008 and in 2009 after 5 years of steady growth. Israel Venture Capital reports showed that in 2009 Israeli

Table 5.3 Entrepreneurial Framework Conditions

EFC1: Financial support
The availability of financial resources, equity, and debt for new and growing firms including grants and subsidies
EFC2: Government policies
The extent to which government policies as reflected in taxes or regulations, are firm size-neutral or encourage new and growing firms. Empirical studies have shown that there are two dimensions or subdivisions of this EFC. The first covers the extent to which new and growing firms are prioritized in government policy generally. The second is about regulation of new and growing firms
EFC3: Government programs
The presence and quality of direct programs to assist new and growing firms at all levels of government (national, regional, municipal)
EFC4: Education and training
The extent to which training in creating or managing small, new, or growing business is incorporated within the educational and training system at all levels. Subsequent empirical studies have shown that there are two distinct subdimensions to this EFC: Primary and secondary school level entrepreneurship education and training and post-school entrepreneurship education and training
EFC5: Research and development transfer
The extent to which national research and development will lead to new commercial opportunities and whether or not these are available for new, small, and growing firms. (The relative level of R&D and estimates of the stock of accumulated knowledge is covered under "Technology" as a General National Framework Condition.)
EFC6: Commercial and professional infrastructure
The presence of commercial, accounting, and other legal services and institutions that allow or promote the emergence of new, small, or growing businesses

Table 5.3 (continued)   Entrepreneurial Framework Conditions

EFC7: Internal market openness
The extent to which commercial arrangements undergo constant change and redeployment as new and growing firms compete and replace existing suppliers, subcontractors, and consultants. Subsequent empirical studies have shown that there are two distinct subdimensions to this EFC: market dynamics, that is, the extent to which markets change dramatically from year to year, and market openness or the extent to which new firms are free to enter existing markets
EFC8: Access to physical infrastructure
Ease of access to available physical resources—communication, utilities, transportation, land, or space—at a price that does not discriminate against new, small, or growing firms. (Presence and quality of these physical resources are covered as a General National Framework Condition.)
EFC9: Cultural and social norms
The extent to which existing social and cultural norms encourage, or do not discourage, individual actions that may lead to new ways of conducting business or economic activities and may, in turn, lead to greater dispersion in personal wealth and income

Source: From Bosma, N. and Levie, J., *Global Entrepreneurship Monitor 2009 Executive Report*, London Business School, London, U.K., 2009.

high-tech companies raised capital at a rate that is 50% below the 2008 rate. At the same time Israeli Angels activity increased slightly to 3.5%, from 3.2% in 2008. The economic global crisis is further reflected in that 50% of young businesses in Israel said that there are less business opportunities than the year before, 65% said that it is more difficult to start new business than a year earlier and 37% estimated that growth possibilities are more difficult (Menipaz et al., 2010).

5.9 Technology Ecosystem*

The technology ecosystem in Israel is affected by economic, education, science, government, and social factors (CIA, 2011). Through both individual

---

* This section is based on CIA reports at http://data.worldbank.org/indicator/IC.BUS. EASE.XQ as well as Government of Israel publications.

Table 5.4 Entrepreneurial Framework Conditions Valued Most Positive (+) and Most Negative (−), per Country

1 Finance
2a National Policy/General Policy
2b National Policy – Regulation
3 Government Programs
4a Education/Primary and Secondary
4b Education/Post-School
5 R&D Transfer
6 Commercial Infrastructure
7a Internal Market/Dynamics
7b Internal Market/Openness
8 Physical Infrastructure
9 Cultural and Social Norms

	1	2a	2b	3	4a	4b	5	6	7a	7b	8	9
Factor-driven economies												
Guatemala		−		−	−	+		+			+	
Jamaica			−		−	+	−				+	+
Saudi Arabia	+			−	−		−		+		+	+
Syria			−	−	−	+	−		+		+	+
Tonga Islands	−			−		+	−		+		+	
Uganda			−		−		−	+	+			+
Venezuela		−	−		−	+			+		+	
Efficiency-driven economies												
Argentina			−		−	+		+			+	
Bosnia and Herzegovina		−	−				−	+			+	

Brazil		−	−		−				+		+	+
Chile		+			−		−			−	+	+
Colombia	−				−	+	−				+	+
Croatia			−		−	+			+	−	+	
Dominican Republic	−				−	+	−				+	+
Ecuador	−				−		−	+			+	
Hungary		−	−		−	+		+			+	
Latvia	−	−				+	−	+		+	+	
Malaysia			−		−				+	−	+	+
Panama	−		+		−		−	+		+	+	
Peru			−		−	+			+		+	+
Russia	−		−	−				+	+		+	
Serbia			−		−	+		+	+	−		
Shen Zhen	−			−	−				+		+	+

(continued)

Table 5.4 (continued) Entrepreneurial Framework Conditions Valued Most Positive (+) and Most Negative (−), per Country

1 Finance
2a National Policy/General Policy
2b National Policy—Regulation
3 Government Programs
4a Education/Primary and Secondary
4b Education/Post-School
5 R&D Transfer
6 Commercial Infrastructure
7a Internal Market/Dynamics
7b Internal Market/Openness
8 Physical Infrastructure
9 Cultural and Social Norms

	1	2a	2b	3	4a	4b	5	6	7a	7b	8	9
Efficiency-driven economies												
South Africa				−	−	+	−	+			+	
Tunisia		+		+	−		−			−	+	
Uruguay				+	−			+	−		+	−
Innovation-driven economies												
Belgium	−		−		−			+	−	+	+	
Denmark				+			−	+	−		+	
Finland		+			−		−	+		−	+	
Germany			−	+	−	−		+	+		+	
Greece			−		−		−	+			+	

Hong Kong		−	+	−	−	+	−	+	+	+
Iceland	−	−		−	+		+	+	+	+
Israel		−	−	−	+	+	+	+	+	+
Italy	−		−	−	+	+	+	+	+	
Netherlands		−		−	+	−	+	+	+	+
Norway	−	+		−			+	−	+	+
Slovenia		−	−	−		+	+	+	+	−
Republic of Korea	+		+	−		−	−	+	+	
Spain	−		+	−		+	+		+	
Switzerland	−			−	+	−	+	−	+	
United Arab Emirates		−		−	−		+	+	+	
United Kingdom		−		−		−	+	+	+	+
United States		−		−	−		+	+	+	+

Source: From Bosma, N. and Levie, J., *Global Entrepreneurship Monitor 2009 Executive Report*, London Business School, London, U.K., 2009.

entrepreneurship and corporate intrapreneurship, Israel's high technology industry has demonstrated a significant growth rate over the past two decades. Israeli high-tech products and services contribute about 10% of GNP and constitute over a 40% of total national export value.* Advanced technologies from Israel in communication, computers, software, medical devices, optics, and consumer goods are in great demand worldwide. Many Israeli developed applications are now imbedded in products of many MNEs. Due to its high technology exports, Israel demonstrates a significant trade surplus. The rate of growth of the technology sector of about 10%–20% annually has sustained itself even during the global economic crisis of 2008–2010.† Government policies have helped nurture the high technology sectors of the economy. "Technology greenhouses" were created by the government, supporting nascent entrepreneurs. These technology greenhouses were largely privatized once proven economically viable. An appropriate tax credit system and subsidies were sanctioned by the government, applied toward larger and smaller technology ventures. High-tech industrial parks around the country were developed, serving as a home for both local and foreign MNEs, such as e-Bay, Microsoft, Google, Motorola, IBM, Intel, and Zoran. Many of these MNEs maintain R&D facilities in Israel, exploiting the local engineering and scientific workforce. Several industries have developed a proficiency in process technology that contributes to their global expansion, such as the case of Teva Pharmaceuticals and Machteshim, a subsidiary of Israel Chemicals (ICL), a world leader in pesticides manufacturing and distribution. At the end of the first decade of the twenty-first century there are about 3000 start-ups, mostly in high technology, the highest per capita in the world.‡

Culture and social norms play a major role in developing an economy based on innovation and entrepreneurial ventures. As can be seen from Table 5.2 it seems that Israel enjoys a relatively high rating in three dimensions: entrepreneurship as a good career choice, a high social status of successful entrepreneurs, and the depth of media attention given to entrepreneurs. Furthermore, the fairly informal relationships commonplace at the workplace and the camaraderie developed during a mandatory 36 months of military service help promote social networking that supports creativity and innovation.

The high technology sector in Israel leverages the freedom of portfolio investment and of FDI and provides employment to some 150,000 workers. It is estimated that Israel spends a significant percentage of its GNP on R&D, at over 4%, compared to the US that spends about 2% of its GNP on R&D. Wages are

---

* Ibid.
† Ibid.
‡ http://www.start-ups.co.il/

relatively low compared to US and Europe, but are significantly higher than the wages in India or China.

As for human resources in support of technology-based industry, Israel's number of engineers and scientists per capita is the highest in the world. They are educated in research universities, such as the Technion-Israel Institute of Technology, Weizmann Institute of Science, Hebrew University in Jerusalem, Tel Aviv University, and Ben Gurion University. For years the government has invested about 10% of its GNP on education, promoting knowledge of sciences and foreign languages. However, there are some concerns about the future. Unlike the 1990s, when immigration from the former Soviet Union kept supplementing the locally educated engineers by supplying talented engineers with modest salary expectations, incoming engineers today come primarily from Israel's academic institutions mentioned earlier, in relatively small quantities. Shortages are mostly in key disciplines, such as software engineering. Also, systemic weaknesses in the education system in Israel and a brain drain could have a negative impact on the future of Israel's high-tech industry.*

Some Israeli companies in the high-tech sector, specifically in telecommunication and software technologies, are already enjoying a worldwide reputation. These are, among others, Check Point Software, M-Systems, Comverse Technology, Amdocs, as noted earlier, and Nice Systems. In the past decades, more than 100 Israeli start-ups have gone public on the NASDAQ stock exchange, while U.S. and European companies have spent tens of billions acquiring Israeli firms. An increase is noted also in high-tech services export, accounting for over 6% of GNP.†

To summarize, the technology intensive sector in Israel relies heavily on an ecosystem consisting of government policies, culture and social norms, investment laws and practices, education facilities, and human resources. The result is indeed the creation of a robust link on a global value chain geared toward R&D, innovation, entrepreneurship, and intrapreneurship. There are many examples in the technology intensive industry that may demonstrate this observation. As a matter of fact it seems that one notices a renewal of the Silk Road through the creation of a virtual global supply chain. As an example, the virtual key board is a result of a technology developed in Israel, financed, partially, by a Hong Kong–based MNE, manufactured in China, and distributed in Europe, North America, and elsewhere. The virtual key board is a complementary product for

---

* For a complete discussion of future concerns see Tadmor, Z., 2011. *The National Science and Technology Policy in Israel*, Haifa, Israel: Samuel Neaman Institute, The Technion—Israel Institute of Technology (In Hebrew), p. 21.

† CIA reports at http://data.worldbank.org/indicator/IC.BUS.EASE.XQ as well as Government of Israel publications.

cellular phones or PDA used primarily in the medical, industrial, and automotive markets. In the medical market, for example, a sterile environment dictates the use of a virtual keyboard, which is a better solution than a plastic cover for the keyboard. The virtual keyboard, which employs laser technology, can be used anywhere, on a train, a plane, a factory floor, or a medical ward. Following a development stage in Israel, it was introduced to Hong Kong–based telecommunication MNE, Hutchison Whampoa. Hutchison Whampoa, through one of its subsidiaries, manufactured the virtual key board in China and launched it globally, starting in the U.K. and continuing in Europe, North America, and elsewhere. Thus, the virtual key board demonstrates the renewal of the Silk Road, creating a supply chain consisting of Israeli technology, Chinese financing resources, and manufacturing capacity and distribution channels in Europe and elsewhere. Indeed, the global interdependence in international trade demonstrates that vibrant economic, political, and technological ecosystems cannot be sustained by operating in isolation. The Silk Road symbolizes that and modern global ventures, such as the one described here, reflect these interdependencies.

5.10 Summary

The ancient Silk Road was an important international trade route that facilitated the trade of goods between East and West, as well as the exchange of cultural, social, and religious values and paradigms. In modern times it has been partially renewed through the efforts of MNEs and governments. By developing appropriate economic, business, transportation, technology, education, political, and social ecosystem, Israel has positioned itself as a relevant and robust link in a global supply chain where R&D, innovation, and entrepreneurship are recognized, encouraged, and supported. Furthermore, it is demonstrated that the national ecosystem of relevance cannot be sustained unless it becomes part of a "Silk Road" where a global supply chain relies on capital and manufacturing capacities sourced globally and global distribution channels leveraged.

Acknowledgments

The partial support provided for this work by The Ira Center for Business, Technology and Society, Ben Gurion University, is hereby acknowledged. Permission to reproduce figures and tables from the GEM 2009 Global Report, which appear here, has been kindly granted by the copyright holders. The GEM is an international consortium and this report was produced from data collected in and received from 54 countries in 2009. Our thanks go to the authors, national teams, researchers, funding bodies, and other contributors who have made this possible. The author wishes to thank the editorial review team members for their helpful and insightful suggestions and comments.

References

Bosma, N. and J. Levie. 2009. *Global Entrepreneurship Monitor 2009 Executive Report*, London, U.K.: London Business School.

Central Intelligence Agency. 2011. *The World Fact Book*, Washington, DC: USA Government. https://www.cia.gov/library/publications/the-world-factbook/

Finger, S. and E. Menipaz. 2008. Regional headquarters of multinational corporations: A new typology. In Ben Gal, I. and Menipaz, E., *Knowledge-Based Competition in the Globalization Era, Digital Proceedings of the 15th National IE&M Conference*, Tel Aviv, Israel: Ortra.

Lowengart, O. and E. Menipaz. 2002. On the marketing of nations: A gap analysis of managers' perceptions. *Journal of Global Marketing*, 15(3/4):65–94.

Menipaz, E., Y. Avrahami, M. Lerner, Y. Hadad, M. Yemini., and D. Barak. 2010. Global entrepreneurship monitor 2009: Observations and trends. In Naveh, E., Ben-Gal, I., Menipaz, E., and Dror, S. (2010). *Getting out of Crisis: Innovation and Competitiveness, Digital Proceedings of the 16th National Industrial Engineering and Management Conference*, Tel Aviv, Israel: Ortra.

Menipaz, E. and A. Menipaz. 2011. *International Business: Theory and Practice*, London, U.K.: Sage Publications.

Miller, T. and K. Holmes. 2011. *2011 Index of Economic Freedom*, Washington, DC: The Heritage Foundation.

Porter, M. E., J. J. Sachs, and J. McArthur. 2002. Executive summary: Competitiveness and stages of economic development. In Porter, M. J., Sachs, P. K., and Cornelius, J. W., The Global Competitiveness Report 2001–2002, New York: Oxford University Press 166–77.

Schwab, C. 2009. *Global Competitiveness Report 2008–2009*, Geneva, Switzerland: World Economic Forum.

Tadmor, Z. 2011. *The National Science and Technology Policy in Israel*, Haifa, Israel: Samuel Ne'eman Institute, The Technion—Israel Institute of Technology (Hebrew).

The World Bank. 2011. *Ease of Doing Business Report 2010–2011*, Washington, DC: The World Bank.

Transparency International. 2010. *Transparency International Annual Report 2010*, Berlin, Germany: Transparency International.

Xu, X. 2003. *The Jews of Kaifeng, China: History, Culture and Religion*, Jersey City, NJ: Ktav Publishing Home.

Yedidsion, H. and E. Menipaz. 2010. Multi national entrepreneurial framework conditions and dynamic career choice: Modeling and analysis. In *Global Entrepreneurship, Innovation and Economic Development, the Fourth Global Entrepreneurship Monitor Research Conference*, London, England, September 30–October 2.

Chapter 6

Decoding Supply Chain Leadership in India

Janat Shah and Debabrata Ghosh

Contents

6.1 Introduction

While the plethora of opportunities offered by India was received with a lot of enthusiasm by corporations across the world, the initial unchecked foray into this country has now turned into cautious planning and execution by the top management of several companies over the last few years (Khanna and Palepu, 2006). Some companies have had success in India, but several others have had to redesign their business models after many years of unmet success (Prahalad and Lieberthal, 2003). Today, organizations realize that the Indian business terrain is far different, and unique, in comparison to the rest of the developed world markets (Khanna et al., 2005). Further, the Indian socioeconomic and political scenarios present unique challenges to businesses and require ingenuity to overcome and ensure success. In such a complex Indian business scenario, where on one hand, the middle class in the country is estimated to be rising, and on the other hand, a significant portion of the population is still earning less than USD 1, businesses have to design unique products and processes to cater to such diverse markets. Organizations that have been able to understand and adapt to the Indian terrain have emerged successful. Out of the several drivers of successful businesses in India, our primary focus is on supply-chain-driven leadership. The importance of supply-chain-driven growth cannot be more emphasized in a country like India where infrastructure problems are plenty.

The question of what makes supply chain leaders different from the average performers in an industry in India has often confounded managers and researchers alike. Through several years of research and discussions with industry practitioners in India, a common underlying observation was found that the leaders "converse" in a different language than their competitors. While the average performers fret over the problems of poor infrastructure, unorganized nature of transportation, warehousing, and long turnaround time at ports in India, supply chain leaders think differently. They focus on creating innovative solutions and, constantly, adapt to the changing business environment of the country. Further our research revealed that supply chain practices of leaders are born out of their business needs, a perspective that the average performing firms often miss out. The leaders maintain their supply chain basics but innovate constantly to counter the challenges that emerge in India. There are several factors that have helped transform these firms into winners.

In what follows, various factors affecting supply chain performance in India are explored. Subsequently, the underlying characteristics of successful firms (supply chain leaders) are discussed.

6.2 Supply Chain Challenges in India

For firms, the challenges of operating in India are numerous and unique to the Indian scenario. Many of these challenges arise due to the economic environment, taxation structures and also the geography of the country.

Varied market structure: India is one of the fastest growing economies of the world, with a growth rate of about 8%–9% in the first decade of twenty-first century. It is fast turning into a large and globally important consumer economy. The Indian middle class is estimated to be 50 million people strong and the Indian per capita purchasing power parity is estimated to significantly increase from 4.7% to 6.1% of the world share by 2015.* This indicates tremendous market potential for companies. However, unlike many economies of the world, the market segment in India is extremely varied with large sections of the population still living under poverty. This also provides various market opportunities. The bottom of pyramid (BOP) (Prahalad, 2004) market in India is estimated to be approximately USD 1.205 trillion, in purchasing power parity terms, which makes up the biggest chunk of the global USD 5-trillion BOP market excluding China. Further, 78% of the BOP population resides in rural India, which makes it challenging for companies to reach out to them. To explain further, in order to reach out to the consumers in rural India, the consumer packaged goods (CPG) sector has to work through smaller pack sizes, which increase the cost of transport manifolds (Shah and Suresh, 2009).

Complex taxation structure: The country offers a difficult tax regime that influences production, warehouse location, and sales aspects of businesses. Often companies are forced to keep one stock point in each state of the country to avoid tax on interstate sales. Also all interstate sales attract a central sales tax. Further, firms have to grapple with state-level taxes like local sales tax, entry tax, octroi, and turnover tax. Such large number of taxes affect supply chain operations. While on one hand, firms deal with this plethora of taxes, on the other hand, several regions in India have established special economic zones that offer tax benefits to companies. In order to benefit from such tax exemptions, several organizations take strategic decisions of locating their facilities in these economic zones. However, facility location decisions when influenced by tax considerations often lead to logistics issues, increased delivery time, restricted market access, and delays in transportation across borders. For example, several pharmaceutical manufacturers have located their facilities at Baddi in Himachal Pradesh (a state in northern India) not because of either market access or resource access, but because Baddi offers taxation benefits (Raghuram and Shah, 2004).

Poor logistics infrastructure: To illustrate another concern, the transportation and logistics sector in India is largely (third-party logistics) an unorganized sector. About 90% of the trucks in the country belong to owners who have less than five trucks. (Also see Chapters 7 and 8 on Turkish 3PL industry.) This unorganized trucking industry often results in unreliable lead times and high transit damages. Also with a large number of old trucks on the road,

---

* IMF, *World Economic Outlook Database*. 2010.

breakdowns are quite frequent, further adding to unreliability. Additionally, customs delays, administrative bottlenecks, long turnaround times at ports, and lack of road infrastructure are other problems in the country. Further, it is estimated that the freight movement in India will increase significantly in the coming decade. This growth cannot be met by the existing railways, roads, and waterways in India unless significant investments are made in the logistics infrastructure. India has seven main road corridors that handle more than 50% of the total annual freight traffic. These corridors are the Delhi–Mumbai, Delhi–Kolkata, Mumbai–Kolkata, Delhi–Chennai, Mumbai–Chennai, Kolkata–Chennai, and Kandla–Kochi. Further, national highways along these corridors handle 40% of road freight traffic even though they are less than 0.5% of the Indian road network. This shows the extensive pressure on the few existing roadways in India. Similarly, rail links on the corridors account for 27% of the Indian rail network but handle over 50% of rail freight traffic in the country. It is further estimated that around USD 45 billion is lost each year due to inefficiencies in India's logistics networks (Gupta et al., 2010). This adds to the inventory holding, wastage, and environmental costs.

Complex distribution networks: The distribution networks in India are largely local and small in scale. Further, distribution systems in India are often left to be developed by businesses themselves like Tatas in trucks and consumer vehicle segment that had to invest significantly to develop their dealers over a long period of time. This also makes it extremely difficult for new entrants in the country to reach out quickly to a large consumer segment. Further, the Indian CPG sector has to work with a very complex distribution system, comprising multiple layers of numerous small retailers, between company and end customer. For example, a company like Marico (Marico is a leading Indian group in consumer products and services in the global beauty and wellness space) has to reach 1.6 million retailers spread (stock keeping units) geographically throughout the country (Shah, 2009). As the numbers of SKUs have been increasing exponentially, just ensuring availability at the last stage of distribution is extremely challenging for companies. In such a scenario, the standard solutions applicable in developed countries are not always suitable for a country like India. Further, institutional voids in India and lack of stringent contract enforcement mechanisms lead to unscrupulous distribution channels and has caused the problem of counterfeit goods.

In the retail markets of the developed world, large departmental/discount chains have managed to grab huge market shares and have clout with CPG companies. Modern retail trade in India is still in an evolutionary phase and accounts for less than 4% of total retail business. Further, margins in distribution are relatively lower in India than the developed markets. Hence, in

India, the CPG manufacturers find it difficult to offer the kind of deep discounts that the modern retailers have been demanding. Further high rental costs in India have made it difficult for modern retail to offer significant discounts. Hence there is not much difference between the price points of convenience stores and modern retail discount stores in India. With this result the Indian market is likely to be dominated by unorganized retail for quite sometime to come.

Skill gaps across sectors: India suffers from a severe lack of trained workforce. The labor force in India is mostly uneducated and lacks the basic training leading to operational delays, accidents, delivery slippages, and health hazards. Among various sectors that get affected due to this, the road freight and warehousing sectors are facing the most severe and immediate need for skill development. Specifically, truck drivers, loading supervisors, warehouse managers, and seafarers need to be developed soon—both in terms of quality and quantity. The most critical skill gap in the road freight segment pertains to truck drivers. The profession attracts primarily illiterate people with no formal training for the job. Most of them start as helpers and over a few years qualify as drivers. It is estimated that nearly 2 million new truck drivers will be needed in the next 10 years, and many modern driving training schools would be required to train them. The warehouse manager position is one of the most affected by the modernization trends in the logistics sector. Existing warehouse managers have typically functioned as administrative heads and commercial managers of small-scale warehouses (such as "godowns" or seaside warehouses) and require a whole new set of skills and extensive training and exposure in modern warehouse operations—including familiarity with warehousing formats, operation of modern equipment, IT systems, industry-specific stocking and handling practices (CPG, perishables, textiles), safety and security procedures, etc. Because of the poor working conditions and relatively less attractive pay and progression incentives, as well as the primarily fragmented and unorganized nature of the industry, logistics in India has suffered from a poor image and lack of attractiveness as a career choice.

Social and political instability: Companies in India often face social and political instability in the country, which leads to supply chain disruptions. For example, in July 2008 nearly 4 million truckers went off the road to protest against rising fuel prices. The agitation led to severe disruptions in food and necessary commodity supplies. In another example the political instability in Andhra Pradesh (a state in southern India) in 2010 has widely affected business centers that had to remain closed in response to political agitations. These disruptions create unique problems for companies operating in India as they need to create risk mitigation plans and be responsive under environments of social and political uncertainty. Often these uncertainties mandate creative solutions from firms and not replication of their global business models. Those who innovate in such circumstances emerge successful in the long run.

To tackle some of these challenges, supply chain leaders in India have devised their own creative solutions. A few common underlying characteristics of such firms have been identified and outlined further.

6.3 Characteristics of Supply Chain Leaders in India

Through various discussions with industry leaders, research, and case studies in India, a few common underlying characteristics of supply chain leaders have been explored.

6.3.1 Supply Chains Driven by Business Needs

Supply chain leaders effectively ensure that their supply chain design and operations are driven by their business strategy. If not, supply chains often lose focus and may turn into liability for firms. For example, Marico's organizational goal of "constant innovation" led it to focus on the introduction of large number of newer brands in its supply chain that would not get adequate attention from dealers because of smaller volumes and uncertain demand. Marico solved this problem by introducing a Vendor Managed Inventory system whose primary objective was to provide greater support to newer brands at the dealer end and not solely reduce inventory. In another example, Amul, which is a Gujarat Cooperative Milk Marketing Federation, came into existence in 1946 to help poor farmers from rural India who had no means of storing excess milk. (Also see Chapter 11 on milk distribution in Pakistan.) The farmers were forced to sell milk through middlemen and had to settle for very low prices. To circumvent the middlemen and improve their returns, a cooperative society was set up in villages of Gujarat (a state in western India). As each village-level society would not have enough volume to justify setting up a milk processing plant, all the village cooperative societies in a district formed a union, which, in turn, collected milk from all the societies and processed it in a centralized processing plant and liquid milk and milk products were marketed to customers all over India. Over the years, Amul has set up a very efficient and effective supply chain that links 2.41 million marginal producers of milk from the rural areas of Gujarat to 500,000 retailers who make Amul products available throughout India (Shah, 2009). The supply chain design and operations of Amul as a cooperative serves the organizational needs of farmers and offers them competitive remunerations today.

The ITC case is another point in favor.* ITC is a private sector company in India with a market capitalization of over USD 30 billion and a turnover of USD 6 billion. ITC entered the branded *Atta* (flour) market with the launch of Aashirvaad Atta in 2002. For ITC, the challenge in the Atta market was to offer consistency in quality. To achieve this, ITC decided to source wheat directly from farmers through its initiative of e-choupal, unlike other players in the market who sourced their wheat from mandis. Through its direct sourcing model the company was able to ensure consistent quality and that's where *Aashirvaad Atta* today scores over its competitors. ITC's supply chain successfully catered to the business need of achieving consistent quality in its product offering and not so much lower cost in sourcing, which might have been the case had its supply chain strategy not been driven by its business strategy.

6.3.2 Design Cost-Effective Innovative Solutions Catering to the Indian Infrastructure Problems

Supply chain leaders uniquely create innovative solutions to overcome the challenges of conducting business in India. Safexpress and the *dabbawalas*† of Mumbai are such cases in point. Safexpress (Safexpress is a leading logistics solutions provider in India) wanted to start a time-definite service for express distribution in the mid-1990s in India (Shah, 2009). Aware of the poor infrastructure and the multiplicity of check points at state borders, Safexpress knew that time-definite delivery across India would prove to be a herculean task. Therefore, before they started the service, they mapped all the routes and identified all the check points and potential areas where trucks may get delayed. They identified 88 delivery gateways, including 44 strategically located hubs so that they can manage time-definite service at an all-India level. Today, its entire fleet is equipped with GPS (global positioning system) units so that any vehicle can be tracked with a precision of 50 m. To ensure that they offer the lowest transit times, they operate 24/7, 365 days a year. Safexpress's key strength lies in its understanding of the business terrain in India.

The world famous dabbawalas of Mumbai came up with an innovative idea to take advantage of the geography of Mumbai and availability of rail infrastructure while simultaneously developing a codification system of the dabbas to achieve six sigma quality supply chains all the while working with poorly educated

---

* http://www.echoupal.com
† *Dabbawalla* is a colloquial term for a person employed in a unique service industry in Mumbai whose primary responsibility is to collect the freshly prepared food in lunch boxes from the homes of the office goers, delivering it to their respective offices and returning the boxes back to the customers' residence.

workforce (Shah, Janat and Ghosh, 2010). The dabbawalas of Mumbai deliver home-prepared food to the middle-class office workers. On every working day, they collect 175,000 lunchboxes (dabbas) from the customers' houses between 7:00 and 9:00 a.m. and deliver the same to the respective offices by 12:30 p.m. The empty lunchboxes are picked up by 3:30 p.m. and returned to the homes of the respective customers by 6:00 p.m. To ensure that no more than one in 6,000,000 deliveries goes astray, the dabbawalas have developed ingenious systems that use a very simple but effective coding system to sort the lunchboxes, on both the forward and the reverse journeys. The extensive use of public infrastructure in Mumbai (local trains) helps keep the operation costs low. The use of local trains and coding system allows them to manage their supply chains remarkably well, which translates into high-quality service, at an affordable cost, for the customer.

To cite another example, meat suppliers in northern India faced challenges in using road transportation to reach the port of Gujarat for exporting meat to Gulf countries as cow slaughtering was banned in the state. They overcame this with the ingenious idea of moving meat through the rail system directly to port (Shah and Suresh, 2009).

Supply chain leaders in India develop local knowledge, understand the cultural mix of the country, and identify seasons of fluctuating demands and plan capacities. Asian Paints (Asian Paints is India's largest paint company and ranked among the top 10 decorative coatings companies in the world), for example, found that in certain districts of Maharashtra (Maharashtra is a state located in the west coast of India) there was a spike in demand for 50–100 mL packs of deep orange shade during a specific period of the year. Further investigations revealed that a few districts of Maharashtra observe a local festival called *Pola*, and during that festival, farmers paint the horns of bullocks with deep orange shade (Shah, 2009). Asian Paints was aware of the fact that the paint-buying decision in India is linked to festivals, and India being a diverse country with different regions celebrating various festivals at different times of the year, it was important for Asian Paints to capture the same in its forecasting models. Detailed analysis of past data allowed Asian Paints to include this aspect in their forecasting process so as to arrive at an accurate demand forecast that enabled them to develop a relevant manufacturing and distribution plan so as to meet the demands of the market place. Poor forecast could have meant either lost opportunity in the market place or excess but imbalanced inventory in the supply chain.

6.3.3 Use of Appropriate Technology

Thirdly, in contrast to popular opinion, for supply chain leaders in India, technology adoption is not a matter of riding on the hype cycles, but rather making a choice driven by their business needs.

For example, for its *Parachute* brand, Marico procures copra (raw material) worth USD 60 million in money value and equivalent to 600 million coconuts in quantity terms annually (Shah, 2009). Copra supply in India has traditionally been in the unorganized market and most of the producers are illiterate. Buying copra on this scale required lot of time and effort on the part of Marico. Marico faced serious challenges in copra procurement because of volatility in prices and uncertain supplies leading to unpredictable returns. In 2004, Marico launched an e-sourcing initiative (implemented in stages), which transformed the buying process gradually from manual to an automated electronic process. Potential vendors today send their quote through SMS and get an electronic confirmation within half an hour. The payment is also made electronically. Through creative use of mobile technology, Marico has managed to bypass intermediaries in its supply chain and reach out to large number of small coconut farmers. Mobile technology has allowed Marico to manage transactions with the farmers in a cost-effective way for offering quotations and confirming orders. Through direct sourcing facilitated by technology, Marico has managed to significantly reduce the variability in its raw material prices and supplies and emerged as a market leader in the hair care business in India.

In another example ITC's e-chaupal initiative leverages the power of information technology to tackle the challenges of Indian agricultural systems characterized by fragmented farms, weak infrastructures and involvement of numerous intermediaries. Within the ITC e-chaupal initiative, village internet kiosks managed by farmers called *sanchalaks* have been set up, which enable the agricultural community to access ready information in their local language on the weather and market prices. ITC in turn disseminates knowledge on scientific farm practices and risk management techniques, facilitates the sale of farm inputs, and purchases farm produce right from the farmers' doorsteps based on information sharing. Transystem Logistics in India, a joint venture between Mitsui and TCI, India, and the logistics partner of Toyota Kirloskar Motors, decided not to implement the expensive GPS systems in its fleet but rather use telecom facilities to track their vehicles once in 4 h. Further, Transystem concentrated on improving the delivery schedules, mapped networks, timings in India and implemented stringent measures to prevent drunken driving, a primary cause of truck accidents in the country. Not only was this solution cost effective but it also improved the reliability and responsiveness of the logistics provider.

6.3.4 Focus on Supply Chain Alignment with Partners

Supply chain leaders in India look beyond the boundaries of their organizations in order to establish long lasting partnerships with their suppliers and dealers.

They align their incentives, their organizational processes, and structures in ways that benefit all the players in the chain. The firms believe in leveraging the capabilities of their partners to help them offer higher product variety with better service offerings. The Bharti Airtel (Bharti Airtel limited is a leading global telecommunications company with operations in 19 countries across Asia and Africa) revenue sharing contract with IBM ensured that Bharti could leverage the capabilities of IBM resulting in higher revenues and profitability for both partners (Shah, 2009). Further in order to monitor the vendor, a dashboard is maintained that shows the key parameters—call drops, blockages, network efficiency, coverage, and capacity among several other metrics. The revenue sharing contract ensures that Bharti's vendors are aligned to work together with Bharti for better customer service. Tata Motors during the launch of Nano had planned for distributed manufacturing where portions of the assembly would take place at the dealer end, thus resulting in lower costs and higher responsiveness. Though Tatas did not implement the idea at this stage, such strategies would definitely be adapted in the future as Tatas try to aggressively manage Nano's cost.

In another example, TVS Motor Company, which is the third largest two-wheeler manufacturer in India and one among the top 10 in the world, has been able to offer 99 different colors to the consumer using modular designs (Shah, 2009). TVS Motors has been offering the Scooty range of two-wheelers for the young generation. In its market research, the company found that color is the prominent way of self-expression among women consumers. Based on this finding, TVS introduced the 99-color campaign in select cities with the intention of attracting young women. The customer can choose from a range of 99 shades, available for a premium of USD 20–40. Offering 99 shades is a supply chain challenge. TVS has come up with an innovative way of managing such a wide variety of offering. TVS stocks unpainted panels at the retail outlet. These unpainted panels are sent to a local vendor who returns the panel, painted in the color chosen by a customer, to the retail outlet within 24 h. So TVS can manage product delivery in 48 h without worrying about the large amount of finished stock at the retail outlets. TVS has achieved success through collaboration with its dealers.

There are plenty of lessons for other firms to follow. Average performers in India often concentrate on their own profits while squeezing their partners by pushing inventory upstream or downstream or forcing suppliers to provide longer credit periods so as to work with negative working capital. Such actions may yield results in the short term but are a death knell for long term business. However, supply chain leaders in India leverage the skills of all participants to maximize performance amidst the unique challenges of operation.

6.3.5 *Global Companies with Local Focus*

Global firms in India sometimes take the easy but wrong step of force fitting technology and processes in India, which exist in their businesses across the globe. However, global brands that have analyzed the Indian terrain in terms of its physical, social, and regulatory challenges and innovatively created solutions have been successful in India. Hindustan Lever (HUL) and Nestle were quick to learn that serving Indian customers in smaller packets via 1 million odd retailers needed very different supply chains than the ones they managed in Europe. Unlike in developed countries, where these companies worked with large pack sizes, in India the trend was in the opposite direction. To increase market penetration in India they needed to reach out to consumers present at the lower end of the economic pyramid. This consumer base could be tapped into only by offering smaller pack sizes. In order to balance market penetration and logistics costs, the two companies successfully adopted the right technology and aligned their internal processes with those of their Indian partners. To illustrate further, HUL initiated a new project called *Shakti* to increase its penetration in rural areas in a cost-effective manner (Shah, 2009). HUL partnered with self-help groups (SHGs) to extend its reach to rural areas, particularly those areas where there are no established HUL distribution networks because of lack of connectivity. A *Shakti* dealer is a member of an SHG, who works as a direct-to-consumer HUL distributor, selling primarily to villages in her neighborhood (Shah, 2009). The business objectives of this initiative were to extend HUL's reach into untapped markets and to develop its brand through local influences. In the process, HUL also provided sustainable livelihood opportunities to underprivileged rural women.

Toyota Kirloskar, which is a joint venture between Toyota Motor Corporation and the Kirloskar Group for the manufacture and sales of Toyota cars in India, works with a fine-tuned transportation strategy for its inbound logistics. At Toyota Kirloskar, the supply received through milk runs from Gurgaon (a large city in the state of Haryana) is cross-docked to trucks going to the Bidadi plant in Karnataka (a state in southwest India). (Also see Chapter 8 on another example of cross-docking operation in Turkey.) Since 2004, Toyota practices double cross-docking, wherein the first cross-docking is done at Gurgaon and the second cross-docking is done at Pune (a large city in the state of Maharashtra), where a vehicle coming from Gurgaon and another from Jabalpur (a city in the state of Madhya Pradesh) are cross-docked so that one big vehicle moves from Pune to Bidadi. Apart from maintaining lower inventory at the plant, cross-docking also helps Toyota in protecting itself from transportation reliability problems. In the traditional system (receipt of FTL shipment directly from each supplier), if the truck in transit is involved in an accident, the entire assembly line will close down for at least a couple of days because of nonavailability of material from one supplier. In the current system, because all the parts from all the suppliers from the north and the west come in on a daily basis, failure of one truck will disrupt supply for a maximum of one day.

Further, successful global players in India have understood that to counter the problem of counterfeiting of goods that is rampant in India, they need to collaborate with their partners. P&G found that various counterfeit products of *Vicks Vaporub* raked in sales equivalent to 54% of the original (Shah, 2009). To prevent such losses P&G exercises greater control over its distribution channels in India and does not just leave it to market forces. Similarly, Microsoft has undertaken significant investments to educate customers and channel partners to curb piracy citing cyber security as one primary reason. It has also started online and phone support to original softwares in India.

6.4 Conclusion

In summary, India offers a huge consumer base to firms across the world and there are plenty of opportunities for business operations. However, the socio-economic and political conditions offer numerous challenges and are constantly changing. Combined with the regulatory framework and taxation regime in India, firms need to be constantly adaptive in managing their supply chain operations. India is showcasing certain trends across sectors, which the firms need to be aware of. In the logistics sector, the global 3PL (third-party logistics) providers, who claim to have the requisite expertise, have entered the market. These new players have to learn a lot about Indian conditions and also are not in a position to offer economies of scale. However, over a period of time, these 3PL companies will develop an understanding of the Indian market and also the relevant capabilities necessary to handle these markets, which will enable them to bring down their costs and to provide cost-effective services to large-volume players. In the taxation scheme, the proposed introduction of Goods and Services Tax will end the distorted tax regimes and lead to convergence of the numerous taxation systems to a uniform tax regime. This will help firms in restructuring their supply chains so as to improve efficiency of their supply chains. Further, the proposed investments in the dedicated freight corridor will reduce the burden on the already stretched road systems and improve overall efficiency in the logistics sector in the country. Firms that constantly innovate to create deep rooted capabilities within their supply chains and leverage them to constantly adapt to such changes, will emerge as leaders in India. As discussed successful firms create these capabilities through development of skills, relationship management with partners, leveraging the power of information technology, understanding the

Indian consumer mindset, and building local knowledge. Further they always ensure that their supply chain practices are born out of their business needs. For firms the learning lies not in blind implementation of global business models in India but understanding and constant adaptation to the Indian business terrain.

References

Gupta, R., H. Mehta, and T. Netzer. 2010. Transforming India's logistics infrastructure. *McKinsey Quarterly*, September:1–72.

Khanna, T. and K. Palepu. 2006. Emerging giants. *Harvard Business Review*, October:60–69.

Khanna T., K. Palepu, and J. Sinha. 2005. Strategies that fit emerging markets. *Harvard Business Review*, 83:6–15.

Prahalad, C.K. 2004. *The Fortune at the Bottom of the Pyramid*. Wharton Publishing, Philadelphia, PA.

Prahalad, C.K. and K. Lieberthal. 2003. The end of corporate imperialism. *Harvard Business Review*, 81(8):109–117.

Raghuram, G. and J. Shah. 2004. Roadmap for logistics excellence: Need to break the unholy equilibrium. Working Paper, Indian Institute of Management Ahmedabad, No. 2004-08-02.

Shah, J. 2009. *Supply Chain Management: Text and Cases*. Pearson Education, Delhi, India.

Shah, J. and D.N. Suresh. 2009. *Global Perspectives-India*. Council of Supply Chain Management Professionals, Lombard, IL, pp.1–48.

Shah, Janat and Ghosh, Debabrata. 2010. Business strategy drives supply chains. *Economic Times*, August: 1–2.

SUPPLY CHAIN PERFORMANCE ON THE SILK ROAD

II

Chapter 7

Borusan Lojistik: Winning in the 3PL Market

Murat Kaya and Çağrı Haksöz

Contents

7.1 Introduction

Stretching across the Anatolian and Thracian peninsulas, Turkey is at the crossroads between Europe and Asia. The Turkish economy has recently been recovering from decades of poor performance. The period between 2002 and 2008 saw an average GDP growth of 6%. A recent special report by *The Economist** mentioned a prediction that "over the next seven years Turkey's growth will match or exceed that of any other big country except China and India." The country is now the world's biggest cement exporter, the second biggest jewelry exporter, and its construction business is in the top three. Boasting the 17th largest nominal GDP in the world, Turkey is cited as one of the leading developing economies, coming behind only the BRIC (Brasil, Russia, India and China) countries.

Turkey is unique in having strong ties with both the Western world (a member of NATO, OECD, European Customs Union, and an official candidate for the European Union) and the Islamic world. The country holds the potential to become a manufacturing center and a transport hub for its region. In fact, Turkey is now Europe's number one manufacturer of TVs and DVD players and number three automaker. Not relying only on the Western European markets, Turkish firms are increasingly doing business with the neighboring countries. Since 2002, the share of exports going to the Middle East countries doubled from 9% to 18%. To support all this growth, the government has been investing heavily in the transportation network, improving highways, railroads, ports, and airports.

Logistics is one of the sectors that benefited most from the economic growth. Between 2002 and 2008, the average growth rate of the industry was 20%–25%.[†] The share of the transportation and logistics sector in Turkey's GDP is estimated to be between 8% and 12%. The country has one of the largest truck fleets of Europe. As of 2009, 1.1 million people are employed in the transportation, communication, and storage services. The logistics performance index (LPI)[‡] of Turkey is ranked 6th among the 24 upper-middle income group countries and 39th worldwide.[§]

The increase in economic activity, coupled with the outsourcing trend, enabled the third-party logistics service providers (3PLs) in Turkey to enjoy steady growth in recent years, only to be interrupted by the recent global recession. The 2008 Turkey logistics industry survey (Quattro Business Consulting,

---

[*] http://www.economist.com/node/17276384

[†] http://www.capital.com.tr/temkinli-buyume-planlari-haberler/20449.aspx

[‡] A measure of logistics friendliness, based on surveys of international operators.

[§] http://info.worldbank.org/etools/tradesurvey/modelb.asp#ranking

2008) claims that the logistics expenditures in Turkey increased threefold from 2002 to 2007, reaching 59 billion dollars. According to the survey, 37% of this market belongs to 3PLs, and the balance belongs to firms' insourced activities. These figures indicate the future growth potential of the 3PL business. As of 2008, 70% of the 3PLs have been in operation for at least 8 years, indicating a certain level of maturity.

The industry is highly fragmented with the top firms (Barsan, Horoz, Omsan, Borusan Lojistik, Ekol, Ceva, Balnak, Reysas, according to 2009 revenue figures*) accounting for only about 10% of the total revenue. Competition is primarily between price and timeliness of service. Progressive customers have started demanding value-added services as well. As of 2010, the industry is still dominated by Turkish firms. However, global capital is moving in, primarily through acquisitions (Investment Support and Promotion Agency of Turkey Report, 2010). Another concern for the industry is the lack of trained labor, particularly white-collar labor.

The current global recession hit the Turkish economy hard, albeit not as hard as most other countries. From 2008 to 2009, Turkish foreign trade decreased by 27%, significantly affecting the logistics industry (Akkaş, 2010). Business slowed down, and customers have been tightening their budgets. Price alone has become the single most important criteria in 3PL selection. In addition to price cuts, customers are also demanding tighter control over their logistics operations. Increased demand uncertainty, volatile exchange rates and fuel prices as well as customer bankruptcy risk have placed immense pressure on 3PLs. In spite of these issues, the logistics industry is fast rebounding from the recession. The revenues of the logistics firms in the Capital 500[†] List have increased 22% from 2009 to 2010; whereas, the revenues of all firms in the list decreased by 7% on average.[‡]

A comparison of two surveys between 2001 and 2007 reveals that the primary concern of the customers of 3PLs in Turkey has switched from service-related aspects such as the timeliness of service to cost-related aspects of the relationship (Aktaş et al., 2010). During the same time period, the rate of complaints about service quality and technical abilities of 3PLs have increased significantly and the customer satisfaction ratio dropped from 40% to 17%. The same survey also found that 67% of the complaints are related to the timeliness of service.

---

* http://www.capital.com.tr/temkinli-buyume-planlari-haberler/20449.aspx
† A list of the largest 500 private firms in Turkey. See http://www.capital500.net/capital/ana. asp
‡ http://www.denizhaber.com/HABER/23796/1/turkiye-39nin-en-buyuk-500-ozel-sir-keti-2010.html

7.2 Borusan Lojistik

Borusan Lojistik (BL) is one of the leading firms in the Turkish 3PL industry. The firm was founded in 1973 to provide logistics services to Borusan group of companies. In 2000, the firm started serving firms outside the Borusan group. In addition to 3PL operations, BL also operates a port in Gemlik, on the southeastern coast of Marmara Sea. The firm achieved a staggering average growth rate of 35% from 2000 to 2007, and 42% in 2008.[*] Due to recession, 2009 revenue was slightly less than that of 2008. Figures 7.1 and 7.2 depict[†] the growth of BL's logistics sales and the increase in the number of warehouses. Figure 7.3 shows[‡] that BL ranks 4th in terms of revenue and 3rd in terms of total warehouse area among the Turkish 3PL firms, as of 2009.

The current CEO, Kaan Gürgenç, came to office in 2002, and spent the first 6 months working on a 5-year strategic plan for the period 2002–2007. The objectives of this plan were successfully achieved: BL completed its national organization, increased revenues fivefold, and started international business. The firm is currently running through its second 5-year strategic plan. The strategic plan is divided into operational plans, and progress is continuously monitored. Mr. Gürgenç's efforts paid off. BL's revenue increased from 35 million USD to 270 million in 2008. What is more, the number of employees fell from 550 to 430 during the same time period, suggesting a phenomenal 10-fold increase in per capita revenue.[§]

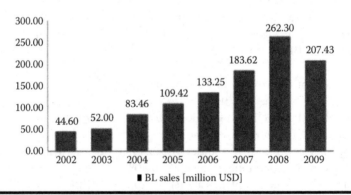

Figure 7.1 BL revenues. (From Borusan Lojistik.)

---

* http://www.lojistikhaber.com/news_sectoral.asp?news_id=2716
† *Source:* Borusan Lojistik.
‡ Compiled from various public resources.
§ http://www.lojistikhaber.com/news.asp?news_id=3591

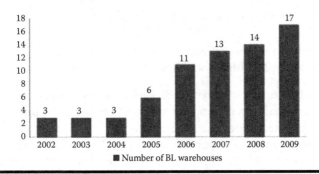

Figure 7.2   Number of BL warehouses. (From Borusan Lojistik.)

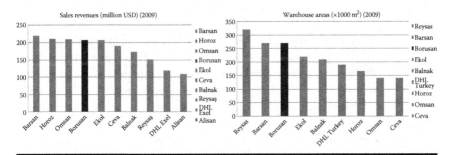

Figure 7.3   Sales revenues and warehouse areas of the leading Turkish 3PL firms. (From Borusan Lojistik.)

Mr. Gürgenç is a firm believer in standards and performance measurement. He made the firm go through ISO 9001, 14001, 18001, and 10002 certification processes, and ISO 20000 and 27001 processes are ongoing. He enjoys monitoring key performance results through a *management cockpit* in his computer. Under the dynamic management of Mr. Gürgenç, BL differentiated itself with its customer and quality focus, as well as its human resource policies and supplier relations.

Customer focus: The Borusan Group, as a whole, is a champion of six sigma methodology in Turkey. BL inherited this attitude. In 2010, BL become the first logistics firm ever to be listed in the "Strategy Execution Champions: The Palladium Balanced Scorecard Hall of Fame" due to successful implementation of Kaplan–Norton balanced scorecard system.* In the ceremony, Dr. Kaplan praised BL for achieving "differentiation through customer-oriented strategies."

---

* http://www.lojistikhaber.com/news.asp?news_id=4716

Since 2003, the firm has operated the "voice of customer" system, a six sigma project to handle customer feedback. In 2007, the firm further restructured its customer feedback mechanism in accordance with ISO 10002 standards, another first in the Turkish logistics industry.

Quality: In Mr. Gürgenç's first years, BL shut down its "quality department" to emphasize that quality is everybody's business.* He made it a priority that the quality standards reach even the remotest offices of the firm. The firm internalized this new management approach quickly and awards followed: In 2006, BL received the EFQM (European Foundation for Quality Management)'s recognition for excellence award. The firm received the "Success Award for Large Companies" in 2008 from KALDER, the Turkish Society for Quality. In 2009, BL's quality standards were crowned by the highly prestigious "Turkish National Quality Award," again from KALDER.

Human resources: BL's human resource practices received the international investors in people (IIP) certificate[†] in 2008. An employee satisfaction survey in 2008 found that 89% of office personnel and 83% of field personnel were satisfied with their jobs, where the Turkish national averages were only 56% and 63% respectively.[‡] Unlike most of its competitors, BL pays more than the minimum wage. Mr. Gürgenç emphasizes the value of BL's people to the firm: "You can buy the best equipment or transfer the best people with money, but you cannot purchase the team spirit."[§]

Supplier relations: BL recognizes the need for collaboration with its suppliers. To this end, in 2006, the firm started a "supplier management system" to reinforce relations with suppliers in different categories. Supplier performance is closely monitored and those achieving superior performance are awarded in an annual supplier gathering. In this gathering, BL also shares its strategic plans with suppliers, and discusses how it expects the suppliers to collaborate in the execution of these plans.[¶] Suppliers prefer working with BL because BL pays promptly and provides good references in the industry. The firm also supports its suppliers' businesses. For instance, BL leverages its information system to find business for independent truckers during their return trips.

BL provides services to select customers from a number of industries mainly from steel, automotive, durable goods, FMCG (Fast Moving Consumer Goods),

---

* http://www.lojistikhaber.com/news_sectoral.asp?news_id=2717
† http://www.investorsinpeople.co.uk/Pages/Home.aspx
‡ http://www.lojistikhaber.com/news.asp?news_id=3591
§ http://www.ambar.com.tr/ambar_lojistik/34806.html
¶ http://www.lojistikhaber.com/news.asp?news_id=5179

1973	1984	2000	2001	2002	2003	2005
BL founded to provide transportation services to Borusan Group companies	First whaf of Borusan Port built	BL restructured itself as an integrated logistics service provider extending its coverage to non-Group companies Second wharf of Borusan Port built	The first owned multi-client warehouse	The first customer warehouse 6 sigma methodology adopted to guide daily work	VOC (Voice of the Customer) management model added to company management systems	LV3 warehouse management software implemented

Figure 7.4   Evolution of BL's service offering. (From Borusan Lojistik.)

construction materials, and chemicals. BL's motto *You just focus on the competition... Logistics is our business* reflects the firm's desire to become a full-fledged logistics service provider to its customers. To this end, the firm has been extending its basic service offerings of transportation and warehousing with value-added services. The evolution of BL's service offerings is summarized in Figure 7.4. While offering new services, BL is also cautious to minimize its operational risks by owning the least amount of assets, and leasing as much as possible (i.e., a light-asset strategy).

We discuss BL's transportation, warehousing, and value-added services in the following three sections of this chapter.

7.3 Transportation Services

Transportation contracts usually have 1–2 year terms. Transportation service does not require significant specific investment by either BL or the customer, and hence, switching costs are low for both parties.

On an average day, BL uses around 650–700 trucks of different sizes (such as pick-up van, light truck, articulated lorry, tankers) and styles (including auto-carriers, chemical carriers, container carriers, steel pipe carriers, cargo carriers). The firm owns only around 120 trucks and provides the rest from an established spot market with a high number of independent trucking companies. BL chooses to own some trucks primarily because this provides some leverage in negotiations with independent trucking companies. In addition, there is the issue of customer perception: Turkish customers somehow trust 3PLs that own some trucks more than those that own none. BL closely monitors the outsourced trucking business to ensure high quality and service to customers.

BL has a good command of the trucking spot market with offices in critical locations, and can obtain good prices for its customers' business. Thanks to such connections, and aided by a transportation management software called Blade,

BL is now capable of arranging trucking operations for new customers within 1 day. However, BL has challenges dealing with the spot trucking capacity as it does not grow as expected, thus increasing the quoted prices. The responsiveness and agility in utilizing the spot market is deemed as a major competitive advantage of the firm.*

7.4 Warehousing Services

Unlike transportation, warehousing relationships require significant investment on both sides. As such, warehousing contracts usually have longer terms, around 3–5 years. In accordance with its light-asset policy, BL does not own any warehouses. Instead, it leases warehouses for 7–10 year terms.

7.4.1 Warehousing Contracting

Typical warehousing contracting process steps are as follows:

1. *The customer sends a request for quotation (RFQ)*: The customer sends an RFQ to a number of qualified 3PLs. The RFQ document explains the customer's business processes in detail and often provides operational data from previous years. The customer defines the terms of the contract and asks for prices regarding 12–15 different items such as
   a. Overnight storage price for a given number of different measurement units (pallets, cases, bags, barrels, steel pipes etc.) per month
   b. Handling fee for a given number of shipments in/out of the warehouse
   c. Overtime service price for shipments in/out of the warehouse
   d. Internal warehouse transfer price
   e. Transfer, storage, and documentation fee for quarantined products
   f. Price for short-term additional capacity allocation in the warehouse
   g. Price for value-added services that the customer may request
2. *BL studies the RFQ and prepares the bid response document*: The 3PLs are given around 2–3 weeks to respond. During this time, BL forms a cross functional team composed of an operations manager, a number of operations specialists, a sales representative and a warehouse manager to study the customer's business and the terms of the RFQ. The team conducts a feasibility analysis by considering the customer's workload requirements

---

* Using slack truck capacity in the logistics supply chain manages the capacity proactively without owning it. See for instance, Hayes et al. (2005, Chapter 3) for more details on this topic. This definitely requires a long tradition of reputation and strength in the market. We believe that BL has competitive advantage on these fronts.

(seasonality, peak periods, etc.) and BL's resource capacities (in particular, workforce and equipment). They run several what-if scenarios to determine BL's optimal bid. The bid prices are tailored for the customer's business. The prices are offered as a menu of choices, and in alternative measurement units such as by square meters, cubic meters, pallets, and racks.

BL teams possess the required industry-specific knowledge for conducting such an analysis.* In fact, BL considers this detailed RFQ study and bid preparation process as one of its competitive advantages. During the process, the team often contacts the customer to receive further information. BL refrains from offering a bid to customers that fail to provide sufficient information at this stage.

3. *The customer collects and analyses bids*: After the bids are collected, some customers demand a detailed cost breakdown from 3PLs to support their bid prices. The customer eliminates candidates that cannot present a feasible breakdown, signaling either questionable financial health or lack of business knowledge. In some cases, to support its prices, BL provides the customer with not only the cost structure, but also with detailed time-study analysis results.

4. *The customer chooses the 3PL*: Price, while always among the main criteria, is not the only important one. One particular customer, for example, evaluates the candidate 3PLs based on 20 criteria including

   a. *Financial health*: Customers often demand audit reports to assess the financial strength of the 3PL during the duration of the contract.

   b. *Information sharing capability*: Information sharing has become so important that some customers ask for detailed contingency plans to be used in case the 3PL's IT infrastructure breaks down.

   c. *Continuous improvement capability*: Customers expect the 3PL to possess domain knowledge and general process improvement abilities. BL's six sigma experience proves invaluable in this respect.

   d. *Capacity to handle future business expansion*: In particular, customers with likely-to-grow businesses demand upward flexibility.

   e. *Communication capability*: Many things may go wrong during a long-run 3PL relationship. When this happens, customers want to face easy-to-communicate personnel. BL people are highly regarded in this respect. They are not only knowledgeable, but also friendly and easygoing.

---

* This industry-specific knowledge should include the intimate knowledge of industry clock speeds since each industry evolves at a different pace. See Fine (1999) for an excellent treatment. Having such know-how and know-why for various industries is crucial for BL to be able to attract and retain more profitable customers in the future.

f. *Adaptability*: Customers with specific value-added service requests question the 3PL's capability to adapt. Having worked with customers from different industries, BL is strong in this aspect of the relationship as well.

The bidding process takes around 2–3 months. If BL is awarded the contract, the team that prepared the bid continues working with the customer. If BL loses, the team is expected to receive feedback from the customer on their performance.

5. *Warehouse setup completed under a letter of intent (LOI)*: BL often rents a new warehouse to handle large customers' business. Finding a suitable warehouse, especially around Istanbul, is a nontrivial task. BL then prepares the warehouse by installing racking systems suitable for the type of product to be stored and by purchasing or leasing equipment such as forklifts and RF (Radio Frequency) scanners. Next, the inventory of the customer is moved into the warehouse. Because the parties often would not like to wait until the legal contract is finalized, these activities are conducted under a LOI, based on mutual trust.

6. *The contract is signed*: The legal contract is usually finalized after the warehouse operation is run for some time period. During this period, previously unforeseen operational details are figured out and the contract is modified accordingly. Being a professional firm, BL always formalizes the relationship with a detailed contract. However, there has never been a warehousing relationship in which BL is involved in legal action based on the contract. The contract is there to stay; yet, it is not the directing force for the business. BL, for example, refrains from pushing customers hard for payments in difficult times. This focus on relationship building is well appreciated by the customers.

The contract specifies estimated levels of activity such as 10,000 pallets staying overnight, or 50 inbound/outbound shipments per day. BL can tolerate deviations of about 15% around these levels. However, further deviations cause significant costs and hence call for renegotiation of the contract terms. Such renegotiation of terms is not uncommon with customers that have fast-pace businesses. It is important for the contracts to allow such flexibility when they are first designed.*

---

* Recently, novel supply chain contract models are proposed that contain flexible options to manage various risks. For example, interested readers are referred to Haksöz and Şimşek (2010) that demonstrates the value of the contract renegotiation option in mitigating breach of contract risk.

7. *The contract is renewed*: Due to mutual hold-up costs, warehouse contracts are more likely to be renewed than transportation contracts. In fact, BL's contract renewal rate is around 95%. The average duration of relationships with key customers is around 4 years, and this figure has been increasing. Yet there have been instances where BL lost customers to competitors that quoted significantly lower prices.

A warehousing contract inevitably introduces risks for BL. This risk, however, is not as high as one might imagine: If the customer fails to use the allotted warehouse capacity, BL might allocate that capacity to some other customer. Yet, to mitigate such a risk, the BL team evaluates RFQs and a customer carefully, that is, conducts its due diligence to weed out potentially risky customers. In addition, for customer-specific warehouses, BL strives to match the duration of the contract with the customer with the duration of the warehouse rental contract.

Next, we discuss an example on how BL is revising its warehousing processes to stay competitive.

7.4.2 Redesigning the Performance Measurement System in Warehouse Operations

During its fast-growth years between 2000 and 2008, BL's primary concern was to capture and retain the business of premium customers. While BL was successful in this rush, it fell behind in improving some of its operational processes. The latest economic slowdown offered a chance to reflect on past performance, and to redesign certain processes in accordance with BL's six sigma approach. Here, we briefly explain how BL redesigned its performance measurement system in warehouse operations.[*]

Reflecting the CEO Mr. Gürgenç's emphasis on measurement, BL calculates a high number of key performance indicators (KPIs) to monitor its performance along many aspects of the warehousing operation. While the set of KPI is well chosen to serve the firm's strategic purposes, the firm realized that it can improve the way the KPI are measured. Actually, BL is not alone in this shortcoming. The State of Logistics Outsourcing 2008 Report cites poor management of KPIs as a major source of concern for customers of 3PLs.[†] As such, performance measurement is not only related to BL's internal purposes, but is also critical for customer service.

---

[*] Interested reader is referred to Zor et al. (2011) for further details on this study.
[†] www.scl.gatech.edu/research/supply.../20083PLReport.pdf

Next, we summarize the issues found in the current performance measurement system and how BL revised the process to address them.

7.4.2.1 Issues in the Current System

A critical examination of the performance measurement system identifies the following issues:

1. *Lack of integration between software systems*: BL uses the warehouse management system (WMS) LV3* to manage orders in the warehouse. Yard operations of trucks, on the other hand, are managed by another software application, the gate tracking system (GTS). GTS was developed in-house by BL to address the yard management shortcomings of LV3. The two software systems operate totally independent of each other, with no connections. This leaves important gaps in the information system. For instance, it is not possible to determine the duration between the generation of a loading order in LV3, and the actual loading time of the order to truck which is stored in GTS.

2. *Lack of automation*: The current system requires manual data entries by operational employees. The employees complain about filling out forms and having to memorize codes. These activities demand valuable time from their already hectic schedule. KPI calculations are also carried out manually at the end of each month. Using MS Excel spreadsheets, the manager of each warehouse calculates the KPI values and enters them into the company-wide ERP system (SAP). This process is not only time consuming, but also prone to error. In addition, top management has to wait until the end of each month to receive the performance report.

3. *Openness to errors and manipulation*: The current system relies on the good faith of the employees. Although BL trusts its employees, it would be desirable to have a system that is immune to human error and potential manipulation. Manual operations and gaps in the current information system leave room for such issues.

4. *Variety of performance requirements*: BL's customers differ greatly with respect to their delivery time requirements. Some require delivery by the same day until, say, 19:00 or 24:00; whereas others are content with next day delivery. There are also customers that demand delivery within a certain time period (for instance within 2, 4, or 9 h) after the order is placed. These differences are not accounted for in the performance calculations of warehouses in the current system.

---

* http://www.la.com.tr/

5. *Different units of measure*: The inventories in warehouses are of different nature, and are measured in different units (US pallets, Euro pallets, cases, bags, barrels, steel pipes, kilo grams, square meters, etc.). The structure of the warehouses is also not homogenous: For instance, some use racks for storage, whereas some do not. These variations make it difficult to measure the flow of goods, or to calculate the warehouse occupancy ratio* consistently between warehouses.

7.4.2.2 How is the Performance Measurement System Redesigned?

BL redesigned the performance measurement system to address the aforementioned issues. In doing so, the firm wanted to achieve quick results with a limited budget, rather than a radical restructuring. The new (redesigned) system is currently in pilot use in Tuzla warehouse. BL will deploy it to all warehouses upon completion of the pilot study, perhaps with required modifications.

The revision attacked the first mentioned issue by establishing a connection between the LV3 and GTS software systems. To this end, the new GTS system prints and attaches a barcode to each and every pallet that enters the warehouse. The barcode matches the pallet to a specific "order code" in the LV3 system. All major steps in the warehousing process (such as receiving, put away, counting, checking, picking, packing, and shipping) are equipped with RF barcode readers. The barcode is scanned at every step, and time information is recorded into a unified database. This approach allows matching of the order flow and the pallet flow, and hence to record the durations of all process steps in the system. Thus, BL solved the first issue by using a familiar and inexpensive technology (barcodes and RF readers), and by modifying the underlying database structure. As a side benefit, this integration will allow the calculation of new KPIs such as *average task completion time* and *average ramp usage time*. Tracking of these measures will further bolster the accountability of the warehouse personnel.

The use of RF barcode scanners also automates the data collection process, addressing the "manual data entry" portion of the second issue. The number of data collection steps is rationalized, and only relevant data is being collected through barcode scans. Manual calculations that were previously conducted with MS Excel or on paper are eliminated. KPI values are also calculated automatically: At the end of each day, the WMS software calculates the KPI values based on the recorded time data, and on each warehouse's specific parameter values that were previously fed into the system. Through SAP, BL's top managers can access these KPIs within the same day. In addition to reducing the time

---

* Defined as the percentage of warehouse capacity being used.

requirements from employees, the automation of data entry and KPI calculations also minimize the potential for human error.

Thanks to automation and the integration of the software systems, the new system actually operates as a management information system (MIS). More than merely supporting performance measurement, the system provides timely "information" to major decision makers, improving visibility for better decisions. The next step in automation will be to integrate this new system with the LAWEP* platform with the purpose of providing the same information to customers. This would definitely improve customer satisfaction levels and allow stronger coordination. Given the increasing information demands of customers, providing such visibility might soon become an "order qualifier" rather than an "order winner." In fact, The State of Logistics Outsourcing 2009 Report[†] found the lack of internal IT integration as the number one complaint of 3PL customers.

Regarding the third issue, the new system is difficult to cheat. A particular employee might try to cheat the new system by scanning the barcode earlier/later to manipulate his performance measure. However, thanks to the recording of all task durations, such a manipulation would affect the duration of adjacent tasks as well, and would easily be caught.

To address the fourth issue, KPI definitions are structured to reflect the needs of different customer and product types. To this end, sales representatives contacted each customer to solidify their definition of "on-time delivery." These updated definitions were fed into the new system to allow automatic calculation.

The fifth issue, on the other hand, turned out to be more difficult to handle. BL gave up on the idea of making cross-warehouse performance comparisons.

7.5 Value-Added Services

Transportation and warehousing are fast becoming commodities. In response, global 3PL firms are increasingly offering value-added services to stay competitive and to enjoy healthy profit margins. UPS, for example, has been offering repair service for Toshiba laptop computers in the USA since 2004.[‡] Under the deal, UPS is responsible for shipping logistics, the repair operation (conducted at a center in Louisville, Kentucky), and parts storage. This arrangement benefits

---

* The web-based supply chain integration software, from the providers of LV3. See http://www.la.com.tr/
† http://www.ro.capgemini.com/en/for_you_to_use/thought_leadership/by_industry/distribution/
‡ http://news.cnet.com/Toshiba-taps-UPS-for-laptop-repairs/2100-1005_3-5201011.html

the Toshiba customers (numbering around five million in 2004) by cutting repair service times to less than 4 days. UPS supply chain solutions web page* offers examples on other interesting value-added services by the company.

BL, too, knows that staying competitive in the long run requires offering value-added services that cater to customers' individual needs. The firm mostly works with international customers. These customers, who are used to obtaining value-added services from 3PLs in their American or European operations, are increasingly demanding such services from BL in Turkey. This requires BL to venture into new tasks and to improve its systems and processes.

The following are some examples of warehousing-related value-added services that BL currently provides:

- Conducting damage check, minor body repairs, and technical testing for import automobiles (Example: In its Köseköy Vehicle Logistics Center)
- Putting gas in the autos' gas tanks
- Providing space in warehouses to customers' technicians for repair operations
- Conducting special packing operations (Example: BL wraps each tire that its customer, a tire manufacturer, sells to Saudi Arabia)
- Labeling/relabeling products
- Preparing product kits and co-packs (Example: A cosmetics company's products)
- Placing warranty documents and user guides in product boxes before shipment (Example: A home appliances manufacturer)

More important than such operational services, BL leverages its experience to become a "solution partner" for its customers. For instance, BL has been operating a tire manufacturer's warehouse since 2009. After collecting data for 6 months, BL determined that the warehouse can be run more efficiently. To this end, BL initiated a lean warehousing project with this tire manufacturer that focused on profiling the distribution of orders over months, simplifying the loading/unloading process and warehouse layout improvements. The project turned out to be a success, praised by the tire manufacturer's managers from headquarters. Ironically, making the customer's warehouse operation more effective may, depending on the contract terms, reduce BL's revenue from the relationship. Nevertheless, BL considers providing such know-how as part of its value proposition to the customer. This definitely helps in forging mutual trust, and opening the way for further collaboration and business.

---

* http://www.ups-scs.com/

BL believes its future customers will be demanding increasingly more value-added services. Next we discuss one such customer example.

7.5.1 Example: A Cosmetics Company's Business

In 2009, BL took over the warehousing and order preparation business of a cosmetics company (CC)* in Turkey. BL reserved part of its 14,000 m² warehouse in Tepeören, Istanbul to CC's business. CC sells cosmetics and beauty products through retailers and pharmacies. Under the contract, CC must forward the orders that it receives from its retail points to BL. BL then prepares the shipment, which typically includes different SKUs, and hands it over to the parcel firm that carries out the distribution operation. This business was the first of its kind for BL, and managers admit that it was quite difficult at the beginning. The business was different for BL in the following three aspects:

1. *Operational difficulties*: The operation is more complex and labor intensive than BL's standard warehousing operations. To handle CC business, BL needs to deal with a high number of SKUs, break bulk pallets, and prepare a large number of small-size shipments quickly. This forced BL to "innovate" a new type of warehouse operation†: To speed up the order picking process, the firm created a "miniature depot" where fast moving items are prepared for pickup. The order is printed using the "Logistics Cashier" module of LAWEP supply chain integration software. Then, the items are collected by hand, and scanned at a computerized checkout counter. This step is not unlike what happens in a supermarket. Finally, the shipment is prepared and handed over to the parcel firm for distribution.
2. *Information requirements*: CC business requires BL to keep and share significant amount of information. BL managers confess that they had to learn a lot of their WMS' (LV3) features, such as detailed order picking, to satisfy CC's information requirements. Information that BL shares with CC includes
   a. Warehouse occupancy rate
   b. Stock levels
   c. Shipment data
   d. Product status (in quarantine, ready, in labeling, etc.)
   e. Order status
3. *Collaboration requirements*: CC business demands a much closer relationship between the firms than what BL is accustomed to. In the beginning

---

* Reference to the cosmetics company is denoted as CC to preserve the confidentiality.
† http://www.la.com.tr/NewsTr.html#

of each month, BL and CC managers come together in an operations meeting where CC managers share their demand forecasts with BL. This helps BL plan for the workforce requirements, which is particularly important before high-demand seasons. To facilitate collaboration, CC opened an office in BL's Tepeören warehouse where CC logistics people are stationed. CC and BL employees become office neighbors, going to lunch together. This physical proximity helps in relationship building, information exchange and allows quick reactions to potential problems.*

7.6 Road Ahead for BL

BL managers believe that the future of their firm lies in the strategic partnerships, epitomized by their CC business. Customers will be demanding more and more customized value-adding services. The 3PLs that can provide these services efficiently in close collaboration with the customers will be the ones that survive in the long run. To this end, effective business procedures and integrated information systems are a prerequisite. BL knows that it needs constant improvement in these areas. Another area that BL may need to consider is the development of more flexible service contracts that allow coordination through risk and information sharing clauses and real options.

BL needs to determine a more focused growth strategy. Until now, the firm has been working with clients from many different industries. This does not create an issue as long as what is provided is transportation and warehousing services. However, providing value-added services is a different game per se where industry-specific experience and scale matters. In the future, BL might benefit from offering an extended scope of value-adding services to a smaller set of customers, preferably from a select set of industries.

The need for an integrated information system will manifest itself more as customers demand real time information in the future. We discussed how BL makes local improvements in its information system to support the new performance measurement process. However, what is needed is a more radical redesign of the whole information system. In particular, BL will benefit most from an integrated logistics suite with warehousing, transportation, port management, contract management, customer relationship management (CRM), service quality management, and web-based collaboration modules.

---

* This collaboration likens to the cocreation knowledge networks seen in Silicon Valley. Innovative ideas are created in a network, which requires close collaboration. Most recently, in an IBM CEO study (IBM, 2010), the most successful organizations turn out to be the ones that cocreate knowledge, products, and services with their customers with an increasing intimacy.

BL should also start developing a "green logistics" strategy. In Turkey, the discussion on the topic so far has been confined to academic circles. BL should champion green logistics in the Turkish logistics industry, not unlike what it did with six sigma, quality, and customer focus. This time, the benefits would reach much larger circles such as governments, public organizations, NGOs, and beyond.

Another area of growing interest to BL is international business. BL has already started doing business in the Netherlands, the USA, Algiers, and United Emirates.* There are many other countries around Turkey (such as Syria, Iraq, Georgia, Azerbaijan, Ukraine, and some Central Asian and Balkan countries), where logistics services are underdeveloped. These developing markets might offer significant business opportunities for 3PLs that can blend world-class logistics operations with strong local connections. In addition to internationals, a large number of Turkish firms in construction, manufacturing, retailing, and service industries have been entering into these markets. Thanks to its stellar reputation at home, BL might be the 3PL of choice for such Turkish firms. This type of strategic move could easily position BL as a pioneer in connecting firms and their supply chains lying on the historic Silk Road.

Acknowledgments

We kindly acknowledge the meticulous research assistance of Ferhat Zor. We are also grateful to Borusan Lojistik executives, particularly Emre Darıcı, Business Analysis and Reporting Manager; İbrahim Dölen, Logistics Director of Turkey; and Özgür Özerbay, Istanbul Anatolian Region Sales and Customer Relations Assistant Manager for their invaluable and generous support throughout the project.

References

Akkaş. 2010. Lojistikte Birleşme Yılı. *Infomag*, Nisan 2010.
Aktaş, E., F. Ülengin, B. Ağaran, and Ş. Önsel. 2010. Türkiye'de Lojistik Alanında Dış Kaynak Kullanımı: 2001'den 2007'ye. 30th YAEM Conference, Sabancı University, Istanbul.
Fine, C.H. 1999. *Clockspeed: Winning Industry Control in the Age of Temporary Advantage*. Basic Books, New York.
Haksöz, Ç. and K.D. Şimşek. 2010. Modeling breach of contract risk through bundled options. *The Journal of Operational Risk*, 5(3):3–20.

---

* http://www.borusanlojistik.com.tr/Hakkimizda.aspx

Hayes, R., G. Pisano, D. Upton, and S. Wheelwright. 2005. *Operations, Strategy, and Technology: Pursuing the Competitive Advantage.* John Wiley & Sons, Hoboken, NJ.

IBM Global Business Services. 2010. *Capitalizing on Complexity: Insights from the Global Chief Executive Study.* May 2010, Somer, New York.

Investment Support and Promotion Agency of Turkey and Deloitte Consulting. 2010. Transportation & Logistics Industry Report. Turkey.

Quattro Business Consulting. 2008. *Turkey Logistics Industry Survey 2008.*

Zor, F., M. Kaya, and Ç. Haksöz. 2011. *Borusan Lojistik: High Speed Performance at the Crossroads,* Sabancı University Teaching Case, Tuzla, Istanbul, Turkey.

Chapter 8

Cross-Docking Insights from a Third-Party Logistics Firm in Turkey

Gürdal Ertek

Contents

8.1 Introduction

In the book *The Silk Road: Two Thousand Years in the Heart of Asia*, Wood (2002) remarkably suggests that "the very name of 'Silk Road' is somewhat misleading." Wood explains as follows: "It [the name 'Silk Road'] suggests a continuous journey, whereas goods were in fact transported by a series of routes, a series of agents, passing through many hands before they reached their ultimate destination."

It's striking to realize that the Silk Road was a complete "cross-docking" operation, executed on a geographically dispersed collection of hubs and routes. Products neither traveled on a single route, nor were they stored on the way. Rather, they were staged for short periods at the hubs and then were passed on toward the next hub on the route, typically by a new caravan or transportation mode.

Today, cross-docking is still very important and popular. Cross-docking has a great potential to bring great financial and time savings in logistics. For example, most of the logistics success of Wal-Mart, the world's leading retailer, is attributed to cross-docking.

In this chapter, different types of cross-docking are reviewed, and the cross-docking applications of a third-party logistics (3PL) firm based in Istanbul, Turkey, are presented. Istanbul was one of the two final destinations on the Silk Road, together with Rome. Today, it is home to the best practices of cross-docking by a multitude of logistics companies, including the company described in this chapter.

Cross-docking is a supply chain strategy that can accomplish significant reductions in both total lead times in a supply chain. In this strategy, cross-dock facilities (CFs) act as transfer points where inbound product flow is synchronized with outbound product flow to essentially eliminate storage of inventory. Two other strategies applied in distribution of products are traditional distribution with warehouses and direct shipment (Simchi-Levi et al., 2007).

In traditional distribution with warehouses, the warehouse typically houses activities of receiving, putting away, storage, replenishment, order picking, and shipping. Storage is well known to contribute greatly to costs due to inventory holding. Order picking is well known to contribute greatly to costs, due to labor requirements or the investments in costly automated equipment (Frazelle, 2001). In pure cross-docking, the activities carried out are receiving, staging, and shipping. Staging of products should last for a very limited time span, such as 24 h, for the facility to be considered a CF. Another essential part of the definition of cross-docking is that the shipments from the suppliers to a CF are preallocated as a whole for a group of customers, even though each unit in the supplier's shipment may not be preallocated.

Increased competition in almost every industry, especially retail and grocery industries has been pushing companies to search for new ways of reducing costs throughout the supply chain. For example, grocery retailers and distributors operate under profit margins of approximately 1.5% (Modern Materials Handling, 2003). Cooperating with supply chain partners to reduce the system-wide costs throughout the supply chain and sharing the benefits is a strategy followed by many companies. The Internet allows companies to communicate among them in real time at costs significantly lower than the past when establishing electronic data interchange (EDI) systems was required for real time communication (Brockmann, 1999). These listed factors have increased the applicability of cross-docking as a supply chain strategy.

Cross-docking in various forms has been in use for a long time, especially by package delivery companies. However, its recent popularity can be attributed to its extensive use by Wal-Mart, which implemented this strategy successfully and eventually became the world's largest retailer with more than 5000 stores throughout the world (Stalk et al., 1992).

Napolitano (2000) provides practical guidelines to planning, designing, and implementing a cross-dock operation. This paper firstly provides a brief tutorial on cross-docking through a review of literature, covering Napolitano (2000) and other sources. Then a case study that describes the cross-docking operations of Ekol Logistics, a leading 3PL firm in Turkey, is presented. The challenges faced by this firm in implementing cross-docking are listed, and insights are summarized.

8.2 Types of Cross-Docking

Napolitano (2000) classifies cross-docking systems into the following three types:

- Type 1 cross-docking: Pre-allocated supplier consolidation
- Type 2 cross-docking: Pre-allocated cross-docking operator (CDO) consolidation
- Type 3 cross-docking: Post-allocated CDO consolidation

When the product is preallocated, its destination is determined at the supplier; when the product is postallocated, its destination is determined at the CF. When supplier consolidation takes place, the supplier builds the final (possibly multi-SKU) pallets that will be shipped to the final destinations. When CDO consolidation takes place, the final pallets are built by the CDO at the CF.

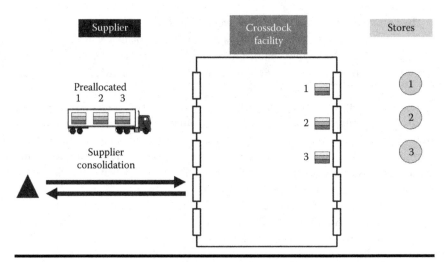

Figure 8.1   Type 1 cross-docking: pre-allocated supplier consolidation.

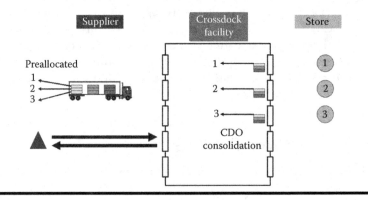

Figure 8.2   Type 2 cross-docking: pre-allocated CDO consolidation.

Figures 8.1 through 8.3 illustrate the three types of cross-docking, as one would encounter in retail industry: Let us assume a supplier that produces three types of product, shown by different colors in the figures. Let us also assume that there are three stores served by the CF, which demand 1/3 of a pallet from each product that the supplier produces. In type 1 cross-docking (Figure 8.1), the three products are consolidated into three pallets, each consisting of 1/3 pallet of each product. The destinations for each of the pallets are preallocated at the supplier. In type 2 cross-docking (Figure 8.2), the destination of each product in each of the three pallets is determined at the supplier; however, they are shipped as single-SKU (stock keeping unit) pallets to the CF. The consolidation into

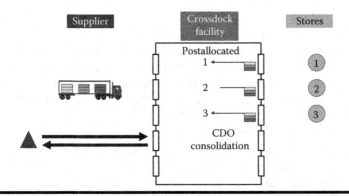

Figure 8.3 Type 3 cross-docking: post-allocated CDO consolidation.

mixed pallets is carried out at the CF, hence the name CDO consolidation. In type 3 cross-docking (Figure 8.3), the supplier sends correct quantities (one pallet of each product) without any label/tag on them that tells their destinations. The allocation of the contents of each pallet to the destinations is determined at the CF, followed by the CDO consolidation.

8.3 Appropriateness of Cross-Docking

Geoffrey Sisko suggests that products with predictable, high demand and high cubic volume flow, and perishable products are ideal candidates for cross-docking (Aichlmayr, 2001). For example, White (1998) reports that the supermarket chain Asda initiated the cross-docking scheme partnering with Kimberly-Clark, the paper industry giant that supplies high-cube (not sure what high-cube means), low-value products such as toilet tissue and paper towels. The choice of these products for the pilot cross-docking program is very appropriate, since these would normally occupy significant warehouse space and cause congestion if traditional warehousing were used (Terreri, 2001).

8.4 Prerequisites of Cross-Docking

The prerequisites of cross-docking, which present certain challenges, can be listed as follows (Napolitano, 2000; Langnau, 2004):

■ *Partnership requirement*: Cross-docking requires total commitment and continuous monitoring at all times by all the parties involved in the cross-docking initiative.

■ *Effective communication between parties*: For cross-docking to operate smoothly information flow has to take place smoothly. This almost always requires investment into information systems technology, and into people that will keep the information systems technology and complex operations working. For example, "Wal-Mart operates a private satellite communication system that sends point-of-sale (POS) data directly to Wal-Mart's 4,000 vendors" (Stalk et al., 1992).

■ *Complexity in managing operations*: The absence of inventories makes it crucial to have a perfect coordination of material flows. Many interrelated decisions at the supply chain and facility level have to be made under numerous resource and time constraints. This is where mathematical models can be of great use.

■ *Sharing the costs and benefits of cross-docking*: Cross-docking may result in savings for some parties and costs or risks for others involved in the supply chain. For example, in a successful cross-docking implementation, the CDO benefits from decreased inventories, labor, and storage space requirements. However, the suppliers involved may have to make significant investment into technology and the retailers may end up with higher inventory levels due to increased lead times (Waller et al., 2006). There should be a complete prior agreement between all the parties on how the costs, savings, and risks resulting from cross-docking will be shared (Kurnia and Johnston, 2001). Another example is the following: The CDO would prefer that the outbound trucks can wait for long time periods such that flexibility is achieved in scheduling the unloading of incoming trucks and the loading of the outbound trucks. However if the trucks are operated by a trucking company, that company would not accept to absorb the cost related with the waiting time of its trucks. Some incentive payment has to be made by the CDO to the trucking company in this case (Schaffer, 1997). Kurnia and Johnston (2001) detail the costs, benefits, and risks associated with each party in a particular supply chain with cross-docking.

■ *Perfect quality requirements*: Suppliers are required to perform perfectly with respect to quality, as inspection has to be significantly reduced at the CF to maintain fast product flow.

8.5 Industries Where Cross-Docking Is Applied

Cross-docking has found extensive applications in retail industry, by companies including Wal-Mart (Stalk et al., 1992), Asda (White, 1998), and Sears (Richardson, 2004). Automotive companies reported to implement cross-docking are Toyota and Mitsubishi (Witt, 1998). Cross-docking is also popular in tele-communications and electronics industries, being implemented by companies such as Ericsson (Cooke, 1999) and National Semiconductor (Richardson, 2004). Another industry that has adapted cross-docking is the apparel industry (Morton, 1996; Shanahan, 2002).

3PL companies, and especially less-than-truckload (LTL) companies, are frequently found to operate under cross-docking. For example, Columbian Logistics, a 3PL company, serves a large grocery wholesaler by consolidating paper products from four large manufacturers, and distributing them to approx-imately 200 stores (Terreri, 2001). Bartholdi and Gue (2000) report cross-docking implementations at LTL trucking companies Southeastern Freight Lines and Viking Freight System. They illustrate an operations research model that was used for determining how to assign inbound/outbound trailers to dock doors at some of the companies' CFs.

As of 2007, 3PL companies in the US were estimated to gross about $110 billion, including value-added warehousing, outsourced carriage, transportation management, freight forwarding, and software (Hoffman, 2007). This under-lines the importance of cross-docking for the industry.

8.6 Benefits and Drawbacks of Cross-Docking

The benefits of cross-docking can be listed as follows (Napolitano, 2000; Aichlmayr, 2001): Cross-docking

- Decreases inventory levels due to elimination of storage
- Enables faster product flow (by eliminating "dwell," the situation of prod-ucts waiting statically at the same location)
- Enables more frequent deliveries
- Enables faster completion of incomplete orders due to more frequent deliv-eries (White, 1998)
- Decreases inventory obsolescence due to reduced inventory and faster product flow
- Decreases labor requirements and costs due to decreased material han-dling (through elimination of put away to storage and order picking)
- Decreases inventory damage costs due to less material handling
- Decreases the amount of space required, and thus increases the handling capacity of the facility
- Supports Just-in-Time (cross-docking is frequently referred to as the "JIT in distribution")

■ Accelerates payments to suppliers (which is an important argument that can be used to convince suppliers to participate in cross-docking)
■ Improves the relations with the supply chain partners

The major drawbacks of cross-docking occur when the prerequisites listed earlier are not met. Other drawbacks, which can be considered as challenges, can be listed as follows:

■ *Stock-out risk*: Since the CF with effectively zero inventory replaces the warehouse with positive inventory, any sudden increases in demand, any unavailability of the product at the suppliers, any delays in the supply chain, or any failure to coordinate perfectly results in a costly stock out.
■ *Union resistance*: The main savings in cross-docking come from decreased inventory and labor costs, where the latter may cause strong resistance among the workforce.

8.7 Implementation of Cross-Docking

At the strategic level, Napolitano (2000) suggests a four-phased framework for making the transition to cross-docking that is composed of assessment and negotiation, planning and design, economic justification, and implementation. It is very crucial that any implementation begins with a pilot program, where cross-docking is initially implemented to cover only a win-or-win subset of products and suppliers. The implementation should then be expanded to include other selected products and suppliers.

At the operational level, the steps involved in a typical retail cross-docking can be listed as follows (Napolitano, 2000; Kurnia and Johnston, 2001; Trunick, 2005):

1. The CDO and the supplier receive order details from the retailer store. If Vendor Managed Inventory (VMI) is implemented, the POS data is sent from the retailer store to the supplier (vendor), instead of the order details, and the supplier initiates a shipment when necessary.
2. If pre-allocated supplier consolidation is carried out, the supplier builds store-specific pallets and labels/tags them. These pallets may be multi-SKU pallets. If CDO consolidation is carried out, then the supplier prepares just single-SKU pallets (to be sorted at the CF). If pre-allocated CDO consolidation is carried out then each case in the pallet should include the information of the specific store it is heading to on a label/tag.
3. The supplier loads the truck that will deliver the shipment to the CF.
4. The supplier sends the Advance Shipping Notice (ASN) to the CDO.

5. The carrier notifies the CDO on the arrival date and time.
6. At the CF, the dock door for inbound receiving is determined and the labor and handling equipment are scheduled to meet the delivery.
7. The dock door for outbound shipment (from the CF) is determined.
8. The outbound carrier is notified of the pickup time, load description, destination, and delivery date and time.
9. The retailer store is notified of the outbound shipment details.
10. The truck/trailer with the supplier's delivery reaches the CF.
11. Manual checks are performed on a small percentage of the supplier's delivery, to ensure accuracy of the ASN.
12. If pre-allocated supplier consolidation is carried out, then the pallets in the inbound shipment are transferred to outbound dock door/truck/trailer. Otherwise pallets are broken into cases. In the case of post-allocated CDO consolidation pallets are allocated to open orders per destination. Then sorted with respect to each retailer store, and loaded to the outbound truck/trailer from the outbound dock door.
13. The outbound truck/trailer leaves the CF and delivers to the retail store.

In Figure 8.4a and b, the steps of a typical type 2 cross-docking operation in retail industry are illustrated. Type 2 cross-docking is selected since Ekol Logistics, the 3PL company that is discussed in this chapter, mostly applies this type of cross-docking. In the figures, the material flows are denoted by black arrows and text, whereas the information flows are denoted by grey arrows and text. All information flows are assumed to take place either electronically over the Internet or an EDI.

8.8 Case Study: Ekol Lojistik

Ekol Lojistik* is a leading 3PL firm and a major CDO in Turkey. The company operates 13 distribution centers (DCs) in İstanbul, Turkey, alone and five other warehouses in other cities in Turkey, with a total warehouse area of more than $285,000\,m^2$ and a workforce of more than 2000 employees.[†] The company manages a fleet of more than 600 trucks in performing its operations.

The main industries that Ekol carries out cross-docking for can be listed as mass retailing (for a client that we will refer to as ABC), pharmaceuticals, and fast moving consumer goods (FMCG). With respect to the classification given earlier in the paper, the type of cross-docking that Ekol implements mostly is

---

* http://www.ekol.com
[†] http://ekol.com/tr/depo-yonetimi/tesisler

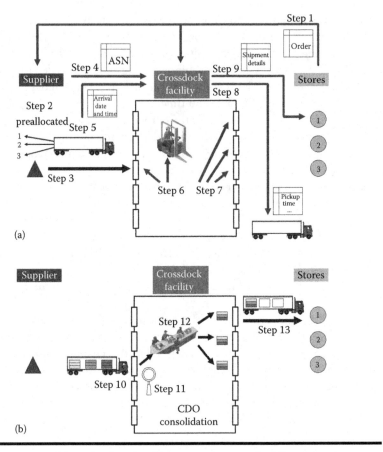

Figure 8.4 (a and b) Steps involved in a type 2 cross-docking operation.

pre-allocated CDO consolidation (type 2 cross-docking), which is referred to as "flow-through" by Ekol managers. The reason that the firm is not able to implement pre-allocated supplier consolidation (type 1 cross-docking) is that most of the suppliers do not wish to undertake the financial and logistic burden of sorting out and labeling their products as pallets before sending them to Ekol's DCs. Thus Ekol undertakes this burden and carries out the sorting, palletizing, and labeling of most of the products that arrive to its facilities. One other reason for implementing type 2 cross-docking, besides the suppliers' reluctance, is the problem of quality that is prevalent for certain manufacturing suppliers.

Pre-allocated supplier consolidation (type 1 cross-docking) takes place only for the products of two major international FMCG companies that are delivered to ABC. This accounts for approximately 30% of the volume that Ekol handles at one of its DCs. Ekol also implements traditional warehousing with put away, storage, and picking for certain products that arrive as a part of the cross-docking activities. These products are separated from the products that are cross-docked and kept for a certain time period until the demand for them is actualized.

Ekol managers also take pride in managing a "project firm." Instead of offering a fixed set of options to clients' requests, Ekol works with clients in analyzing their supply chains with respect to many dimensions and determining a customized solution. For example, Ekol works with data supplied by clients to compute the increase in costs and lead times if pre-allocated CDO consolidation (type 2 cross-docking) is carried out instead of pre-allocated supplier consolidation (type 1 cross-docking). Ekol also quantifies the increase in costs and lead time when traditional warehousing is carried out instead of cross-docking. The increases in costs and lead times depend heavily on the industry, client, product, supply chain, and market characteristics.

One of the basic reasons that ABC considered outsourcing its logistics operations to Ekol was to eliminate the long truck queues that used to accumulate in front of ABC retail stores. These trucks used to arrive from a multitude of suppliers to deliver LTL quantities. In the logistics activities that Ekol executes, the suppliers' trucks arrive at Ekol's DC and unload their (mostly non-palletized) loads. Ekol then consolidates these products into pallets and ships them to ABC stores immediately.

Ekol faces many challenges in planning and executing the cross-docking operations for ABC and other clients. Most of these challenges are in fact valid for traditional warehousing as well; however, due to stricter time constraints they are more heavily pronounced in cross-docking operations. Some of these challenges and the solutions employed by Ekol can be listed as follows:

Receiving non-palletized shipments: Due to the lack of transportation conditions in Turkey, Ekol typically receives non-palletized shipments. This necessitates stricter quality control in receiving operations, and more workers for sorting for the cross-docking operations.

Meeting the delivery requirements: The shipments out of the DCs are almost always unidirectional, that is, they involve only one way shipments to retail stores. However, especially given the very high fuel prices in Turkey (which are approximately three times those in the United States) the revenues would not break even with the costs if Ekol used only its own dedicated fleet. Thus Ekol purchases transportation service from trucking companies. The trucking company for a particular outbound shipment is selected within hours based on whether it has a shipment for the return trip (so that the trucking company will charge only for the delivery trip).

Assuring delivery quality: To assure quality in delivering shipments to retail stores, Ekol prefers to work with a selected group of the best-performing trucking firms on a regular basis.

Delayed deliveries: Some suppliers deliver their products with delay. In this case their delivery to destinations is delayed to the next shipment.

The lack of planning by some of the clients: Ekol DCs can become extremely crowded if the clients do not plan their operations properly. For example, the two products that have to be matched for shipping may be arriving in distant time periods, even though they have to arrive in close proximity to match them rapidly and thus carry out cross-docking. The solution that Ekol implements is charging the clients not based on volume alone, but also based on the DC space that they occupy. This indirectly disciplines the suppliers to send their shipments so that they can be coordinated with outbound flow and other inbound flow.

Facility limitations: Land is very scarce and extremely expensive in Istanbul, Turkey. This makes it difficult to find land to build new DCs that will serve cross-docking. Also, almost none of the existing DCs are built for cross-docking in mind at the first place. Ekol thus has to adopt its operations into existing facilities. For example, the cross-docking operations for ABC are carried out at a DC recently acquired from a furniture producer firm. This DC contains docks on only one side of the building, so cross-docking has to be carried out only from one side of the building. However, a DC with dock doors on both (or even all) sides of the building could have been more efficient. Ekol tries to resolve this problem by enforcing standards when renting/buying a semifinished building that will serve as a DC.

Seasonality in products: Seasonality exists in many product families and products. Ekol alleviates the load that this would put in its distribution operations by diversifying its client portfolio such that the demands of different clients complement each other and the flows are balanced throughout the year for all operations.

Quality concerns: Some inbound materials (coming especially from certain countries) to the Ekol DCs have to pass from a more strict quality control. This increases requirements for labor and floor space.

Customs regulations: The customs limit some of the operations of Ekol. For example, sportswear other than sports shoes pass through Halkalı Customs Office, whereas sports shoes are categorized as shoes and pass through Tuzla Customs Office. Even though both of these customs offices are located in İstanbul, Turkey, there is a distance of 75 km between them and a great difference in their distance to each of the Ekol DCs. This situation only increases the coordination burden of Ekol, since now the same demand points may have to be served from two different DCs for two different products.

Traffic regulations: Cross-docking requires fast loading and unloading of materials at the DC docks. To enable this, Ekol prefers using tail lifts at the back of the vehicles that connect to the docks and speed material loading and unloading. This equipment typically weighs between 750 kg and 1 ton. However, traffic regulations already limit the weight of the load of a vehicle, so the tail lift equipment limits the capacity of the truck even further. Ekol resolves this issue by utilizing the vehicles with tail lifts for carrying only low-density products.

   One of the greatest challenges for Ekol in planning cross-docking operations is the very short time span available for decision making. Ekol manages its operations with the help of a warehouse management software (WMS) developed in-house. During the interviews with the Ekol managers, the author has noticed that some of the decisions can be made much faster and probably more efficiently through use of appropriate decision support systems. One of these decisions is the problem of loading the vehicles efficiently under various constraints. These constraints include assuring that

- The food items are not loaded next to nonfood items
- The pallet heights do not exceed 2.2 m
- The pallets are loaded on top of each other such as to avoid crushing
- An SKU is kept in the smallest number of pallets possible

While keeping ABC as a major customer for more than 5 years, the cross-docking operations of Ekol Lojistik has also brought new customers to the company in diverse industries, and has also contributed to the company's reputation overall. Ekol Lojistik has opened 5 of its 15 current warehouses within the last 5 years, and has nearly doubled its closed warehouse area. Thanks to its cross-docking strategy and supply chain innovations, in recent years the company has expanded its operations to neighboring countries, as well as the Scandinavian corner of Europe.*

8.9 Conclusions

In this chapter, the types of cross-docking were identified; the situations and industries where cross-docking is applicable were explained; prerequisites, advantages, and drawbacks were listed; and implementation issues were discussed.

---

* http://ekol.com/tr/kurumsal/kilometre-taslari

Finally a case study that describes the cross-docking applications of a 3PL firm in Turkey was presented, and the challenges faced by the firm and the remedies to these challenges were explained. Since cross-docking requires decision making in compressed time intervals, there is a potential for the application of mathematical models and a decision support system that enables the user to make the best decisions in short time intervals.

Acknowledgments

The author is thankful to Can Kircan, Cem Kumuk, and Alper Hubar of Ekol Logistics for allocating their time for the interviews. The author would also like to thank Professors Beverly Wagner and René de Koster for their valuable suggestions. A precursory work related to this chapter has appeared in the International Logistics and Supply Chain Congress 2005, İstanbul, Turkey, organized by Galatasaray University, LODER (Logistics Association Turkey) and Universite de Paris-1. The author is thankful to congress organizers Professors Gulcin Buyukozkan and Mehmet Tanyas for their suggestions.

References

Aichlmayr, M. 2001. Never touching the floor. *Transportation & Distribution*, September 2001, 42(9):47–52.

Bartholdi, J.J. and K.R. Gue. 2000. Reducing labor costs in an LTL crossdocking terminal. *Operations Research*, 48(6):823–832.

Brockmann, T. 1999. 21 warehousing trends. *IIE Solutions*, 31(7):36–40.

Cooke, J.A. 1999. Making the global connection. *Logistics Management and Distribution Report*, June 1999, pp. 47–49.

Frazelle, E.H. 2001. *World-Class Warehousing and Material Handling*, McGraw-Hill, New York.

Hoffman, W. 2007. 3PLs reach record revenue. *Traffic World*, 23:18–19.

Kurnia, S. and R.B. Johnston. 2001. Adoption of efficient consumer response: The issue of mutuality. *Supply Chain Management: An International Journal*, 6(5):230–241.

Langnau, L. 2004. Crossdocking comes of age. *Material Handling Management*, July 1, 2004, 59(7):49–53.

Modern Materials Handling. 2003. Grocery distribution the squeeze is on. *Modern Materials Handling*, September 2003, G3–G12.

Morton, R. 1996. Design customer service into your space. *Transportation & Distribution*, May 1996, 37(5):124–129.

Napolitano, M. 2000. *Making the Move to Cross Docking*, Warehousing Education and Research Council, Oak Brook, IL.

Richardson, H.L. 2004. Execution at the dock. *Logistics Today*, 45(4):31–33.

Schaffer, B. 1997. Implementing a successful crossdocking operation. *IIE Solutions*, October 1997, 29(10):34–36.

Shanahan, J. 2002. Cross docking spruces up Urban Outfitters. *Logistics Management*, 43(1):65.

Simchi-Levi, D., P. Kaminsky, and E. Simchi-Levi. 2007. *Designing and Managing the Supply Chain: Concepts, Strategies, and Case Studies*, McGraw-Hill/Irwin, Boston, MA.

Stalk, G., P. Evans, and L.E. Shulman. 1992. Competing on capabilities: The new rules of corporate strategy. *Harvard Business Review*, March–April 1992, 70(2):57–69.

Terreri, A. 2001. Profiting from cross docking. *Warehousing Management*, September 2001, pp. 29–34.

Trunick, P.A. 2005. Time is inventory, *Logistics Today*, April 2005, pp. 26–27.

Waller, M.A., C.R. Cassady, and J. Ozment. 2006. Impact of cross-docking on inventory in a decentralized retail supply chain. *Transportation Research Part E: Logistics and Transportation Review*, 42(5):359–382.

White, D. 1998. Asda floats shipshape supply plan. *Supply Management*, April 9, 1998, 3(8):15.

Witt, C. 1998. Crossdocking: Concepts demand choice. *Material Handling Engineering*, July 1998, 53(7):44–49.

Wood, F. 2002. *The Silk Road: Two Thousand Years in the Heart of Asia*. University of California Press, Berkeley, CA, p. 9.

Chapter 9

Balance of Power between Buyer and Supplier: The Case of Chinese and Western Companies

Oliver Schneider, Robert Alard, and Josef Oehmen

Contents

9.1 Introduction: The Appeal of China and the Importance of Power in Buyer–Supplier Relationships

Being almost 3000 years old, the Silk Road is probably the oldest monument to the globalization of trade. Although it encompasses a region much bigger than Western Europe and China, the trade between those two areas is the focus of this chapter. In particular, we analyze the balance of power between European buyers and Chinese suppliers. A number of reasons make the Chinese procurement market attractive, some of them for thousands of years:

■ *Availability of technologies and capacities*: There exist special cluster regions in China (leading areas for some key industries, e.g., textiles or magnets). This results in a necessity to source from China, as key suppliers for these products and technologies are available only there. Historical examples include the production of silk, tea, or porcelain that were either not available at all outside China or available only in much lower quality.

The incentives that became more dominant in modern times to procure goods from China are:

■ *Cost*: Buyers aim for cost reductions, mainly due to the lower labor and regulatory cost.
■ *Sales*: To get an access through the supplier to the local (Chinese) market.
■ *Chinese local content requirements*: Sometimes there is a demand from the customer side, either due to business development decisions or Chinese regulatory requirements, to set up a local production or supplier base in China.
■ *Tax breaks*: Companies aim for global tax optimization in complex manufacturing and trade networks.

The interface between Western companies and the Chinese suppliers is often complex. Still, many barriers have to be overcome, for example, logistics, IT integration, language and cultural barriers, technical and business norms, or intellectual property rights (IPR). Another specific problem is the substantial loss of bargaining power in the buyer–supplier relationship over several years leading to disadvantages of the Western companies (Oehmen et al., 2009a). As a result, the Western buyer cannot exert enough influence on the supplier's performance in terms of quality and delivery, as well as, the price of the goods anymore. The supplier might not extend or even cancel the contracts with the Western company. One main reason behind this issue is the high growth rate of Chinese companies due to the strong economic environment in China. The Chinese suppliers are often growing much faster than their Western counterpart, so that the importance (i.e., relative purchasing volumes) of the Western customer is decreasing from the Chinese supplier's perspective.

This topic is also very relevant to Chinese companies. Due to the increasing "bad" experience that some Western companies had with long-term relationships in China, there is a growing reluctance to make long-term commitments. This has a direct influence on the Western companies' willingness to develop Chinese suppliers, for example, through technology transfer or know-how transfer. But Chinese companies that are interested not only in quantitative, but also qualitative growth, have a strong interest in encouraging these long-term relationships.

Sourcing relationships and their management are generally discussed to some extent in literature. For instance, Kraljic (1983) introduces a structuring method for sourcing items and suppliers, and the basic strategies and measures that can be derived from the different combinations. Caniëls and Gelderman (2005, 2007) analyze the effect of different strategies on the power in buyer–supplier relationship. They also develop a power and dependence portfolio that extends the portfolio as introduced by Bensaou (1999), which is discussed later in this chapter. Nyaga et al. (2010) introduce a comprehensive governance model for collaborative partnerships. This model provides a perspective on how collaborative activities, via key mediating variables, influence the relationship outcome (i.e., performance). Another model was recently introduced by Autry and Golicic (2010), who describe the "relationship strength-performance spiral." The spiral relates the buyer–supplier relationship strength to its performance, whereas the relationship strength is influenced by past performance experiences and the changing expectations during the course of time. It includes decision points at which one party can decide whether to amplify the relationship performance through investing in the partnership, or to let the spiral correct itself downward until the termination of the partnership. However, all of the authors implicitly assume partnerships with similar growth rates. But, as this chapter will show, sourcing relationships need to be managed with a broader perspective, by, among other things, taking changing bargaining power into account, which can be caused by several reasons.

With regard to the extremely dynamic Chinese industry environment, it can be said that currently the available literature does not provide enough practical and actionable hints for an effective management of sourcing relationships in fast developing countries. This situation is even stressed by the big cultural differences between Western and Chinese supply chain partners. Therefore, the purpose of this chapter is to provide (1) a perspective on the development and the resulting implications of the balance of power between Western and Chinese companies and (2) a list of concrete actions to facilitate sustainable relationships with Chinese suppliers.

9.2 Research Method: Background to the Findings on the Balance of Power

Within two Swiss funded research projects* the topics were investigated via case studies of 16 buyer–supplier relationships of 14 Western and Chinese companies, operating and/or located in China, to gain an understanding from both perspectives. The case studies answer the following questions:

1. How did the buyer–supplier relationships develop over the years (bargaining power, power balance, dependence of the supplier, and the dependence of the buyer)?
2. What are the key reasons, causes, and indicators for this development?
3. What are the view and the consequences of these developments from the Western and the Chinese perspective in particular?
4. What measures can be taken to develop and sustain a mutually beneficial and stable long-term buyer–supplier relationship?

Against this background, an interview questionnaire was developed. The interviewees had to describe some typical relationships with regard to the development of the power relationship and the implications on the collaboration.

The analysis was limited to relationships with suppliers who produce the sourced good itself. Also, the analysis was limited to relationships where only one type of good is sourced from the supplier, in order to be able to clearly and easily characterize the relationship and the influence of the complexity of the product. In order to characterize the power relationship and identify measures to change the power situation, criteria and examples based on Oehmen et al. (2009) were provided in the interviews. The case studies were categorized according to the procurement object, its characteristics, and the sizes of buying and supplying company.

---

* Project "DC-SC-M", Swiss commission for technology and innovation CTI, and project "Risk Minimization in Global Sourcing", Swiss national fund (SNF).

9.3 Fundamentals of Power in Buyer–Supplier Relationships

In the context of this chapter, power is defined as the relative difference in the mutual dependence of buyer and supplier (Bacharach and Lawler, 1981). Criteria for assessing a dependency are provided in Table 9.1. The criteria are assessed subjectively on a free scale between weak and strong from both perspectives. If any of these criteria is assessed as a strong dependence by one party, the total assessment of the dependency will be usually considered as a strong dependence, as one criterion is in most cases enough to put one side into a strong position of power.

After both sides assessed the dependency situation from their perspective, the result can be displayed in a 2 × 2 portfolio, following the logic described in Bensaou (1999) and shown later in Figure 9.1. The y-axis displays the dependence of the supplier, whereas the x-axis depicts the dependence of the buying company. When both sides rate their dependence as low, the power situation is

Table 9.1   Factors for Assessing the Dependency of Buyer and Supplier

Product-related factors	Potential for substitution of the procurement object
	Relative share of sales or purchasing volume
	Profit contribution of product
Company-related factors	Capital situation
	Amount of specific investments into particular relationship
	Potential for threatening backward or forward vertical integration
	Transparency about the situation of the other party and industry
Industry-related factors	Level of concentration in the industry
	Market entry barriers in the industry of the other party
	Costs for substituting the other party
	Aggressive rivalry within the industry
	Importance on the market

Sources: From Oehmen, J. et al., Risk minimization in global sourcing by managing the bargaining power in buyer–supplier relationships, in *Proceedings of Third International Conference on Changeable, Agile, Reconfigurable and Virtual Production (CARV)*, Munich, Germany; Oehmen, J. et al., *Ind. Manag.*, 25, 29, 2009.

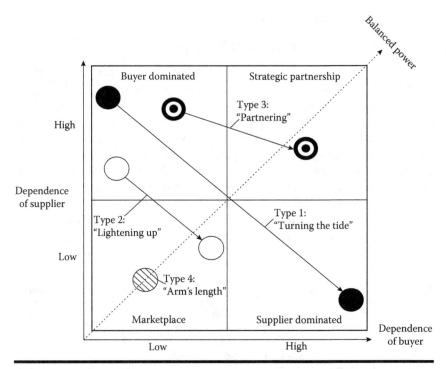

Figure 9.1 Typical developments of power relationships found in the case studies.

classified as a typical marketplace scenario. When both consider the dependence as high, the relationship is a strategic partnership. Only in cases where the supplier (buyer) assesses its dependence as high, but the counterpart as low, the relationship is considered buyer (supplier) dominated.

When it comes to deciding on whether to continue or terminate the relationship, including the negotiation on the collaboration conditions, the assessment of the current power situation also provides a perspective on the future of the bargaining power of the two parties.

9.4 Examples: Typical Developments of the Balance of Power

The objective of the following examples is to provide a perspective on the development of buyer–supplier relationships, the reasons behind these developments, and the resulting consequences and to provide insights on how to establish stable and long-term relationships. Against this background, interviews with nine Western European and five Chinese companies were conducted, nine of them in China, and five in Western Europe. One of the Western companies, as well as one of the Chinese companies, provided 2 different cases, leading to an overall number of 16 case studies. The sizes of the companies differ between 10 and 100,000 employees, with turnovers between USD 2 million and USD 30,000 million. The case studies cover the industries of mechanical engineering, precision technology, electronics, textile, automotive, and renewable energy. The complexity of the sourced products ranges from simple machined parts over commodity electronics to complex cast iron.

The interviewees were asked about the initial power situation in the selected relationship, the current situation, and the reasons for the development. The interviews concluded with a discussion on what could or should have happened ideally, in order to create a win-win situation. The developments of the 16 case studies were clustered to 4 generic types, which are shown in Figure 9.1.

Within the interviews, all the described examples started with a buyer-dominated relationship. However, the interviewees confirmed that they also have many relationships that are considered as "marketplace" or "arm's length," without significant changes of that situation (type 4). In the following, this chapter focuses on the relationships that showed a dramatic change (types 1–3), in order to provide insights on how to avoid negative developments and achieve a mutually beneficial situation.

9.5 Explanation: Reasons for the Development of the Power Situation

The case studies support the notion that Western companies often lose bargaining power during the course of the relationship with the Chinese suppliers. For this development, the following reasons are given in the case studies:

■ *Change of the (relative) market situation of the buying company*: In cases following type 1 ("Turning the Tide") in the portfolio, the Western buyer begins the relationship from a very strong position, but ends up depending on the supplier. In the 1st years of the partnership, the buyer can dictate prices and payment conditions. But because of the high buying volume, the relationship is still profitable for the Chinese supplier. Several cases showed a situation where the sales volumes of the buyers decreased heavily as they experienced difficulties in their main sales markets. This in turn resulted in a much lower buying volume toward the supplier. In all these cases the supplier was not anymore as dependent on the buyer as in the beginning, because of other, also local, customers the supplier acquired in the mean time. The reduced order volumes further reduced the importance of the buyer for the supplier. At the same time, the buyers became highly dependent on the supplier, whom they often needed to keep supplying them at favorable prices in spite of their own reduced influence to continue their business.

■ *Different growth rates*: Relationships following type 2 ("Lightening up") start buyer-dominated, with the Chinese supplier being relatively small, but experiencing tremendous growth rates and acquiring many new customers. The Western buyer shows much lower growth rates. The supplier values the buyer because it gains know how and builds up its reputation in the market. However, after several years, the buyer is only one out of many, accounting for only a small share of the supplier's sales, and without any apparent effort to maintain or improve the quality of the partnership. This results in the situation that the buyer loses more and more negotiation power over time, sometimes without even being aware of it.

■ *Investment in the partnership*: In cases following type 3 ("Partnering"), the relationship is often initiated by a Western company looking for a cheap Chinese manufacturer of standard (norm) parts. The Western buyer builds up the supplier with a transfer of know how. With this support the supplier is able to acquire other customers, becoming more independent of the original buyer. Being aware of that situation, the Western buyer approaches the supplier with the question what other parts could be part of collaboration, increasing not only the buying volume quantitatively but also qualitatively. For instance, the product range can be widened to specific and customized parts. In most of the cases the buyer again supports the supplier in setting up the more complex production.

Please note that the previous classification is idealized. The described influencing factors within a specific development type can often be also found in other types. However, the relative importance of the individual influencing factors determines the direction of the power relationship development within the portfolio.

9.6 Suppliers' Perspective: Perception of and Consequences for Chinese Suppliers

When interviewing the Chinese suppliers in the case studies, questions about their perception were added, in order to gain a better understanding of cultural differences. The results can be summarized as follows:

■ *Trust and default risk*: In one case of type 1 the capital situation of the Western buyer became critical during the partnership. Because the Western company did not communicate openly about its economic situation, the Chinese supplier got doubts about the quality of the partnership. These doubts regard quality in terms of *trust*—"Why don't they tell us about problems; we could deal with them cooperatively?"—and *default risk*—"Is it safe to conduct business with this partner; will I get my money?" As a consequence, when buying volumes dropped, the Chinese supplier asked for higher prices, in order to remain profitable. When rumors about bankruptcy appeared, it asked for advance payments. The buyer had to consent to both requests. However, the Chinese supplier stressed that it would have behaved differently if its perception on the partnership quality would have been better.

■ *Give and receive*: Relationships following type 3 are perceived as ideal realizations of partnership thinking. In this scenario the buyer constantly considers what is more profitable for the Chinese supplier. For instance, in one particular case study both companies were moving in a highly standardized business and faced the same challenge with low profit margins. So a new product area was introduced, allowing both sides to grow together and realize a higher profit. As a consequence, the supplier put a higher priority on the buyer's orders, although other customers had higher order volumes.

■ *Imbalance of investments*: In another type 1 scenario the supplier highly appreciated the good start of the relationship and the initial investments of the buyer in terms of know-how transfer. However, it would have expected that the buyer continuously invested in the relationship, in order to grow together. Because this did not happen, the supplier saw no prospect for the future and continuously reduced the sales commitments, in order to free capacities for other customers.

A major general finding of the interviews is that Chinese suppliers only break the relationship when they have a bad perception on its quality in terms of a partnership attitude. Chinese suppliers are willing to continue business relationships also when the relative share of buying volume is small, as long as the relationship is perceived as long-term oriented and stable partnership based on trust. The Chinese term and concept of *Guanxi* was mentioned several times, understood as the "special relationship" between two parties (Fan, 2002). For instance, in the last described case, the interviewee stressed that he would have preferred a development according to type 2, because this would have been a better realization of Guanxi principles.

Although Guanxi was already recognized as a very important difference to Western business cultures in previous studies, there still seems to be a big gap in the theoretical recognition of Guanxi and actual business behavior of Western companies.

9.7 Taking Action: Measures for an Ideal Development of the Relationship

During the interviews, best practices and advises for a more sustainable relationship with Chinese suppliers were collected. The measures are focusing on developing a strategic partnership with the Chinese supplier, or at least a marketplace relationship, as this development was considered as best by the interviewees. A selection of the most important practical measures for the Western (buying) company is the following:

- *Communicate openly*: Chinese suppliers want to be able to plan their profits. Due to this, they want safe relationships based on trust. An open communication about the capital situation is highly appreciated. One interviewee even mentioned that an involvement of the supplier in terms of finding a solution for a liquidity problem would be an appropriate measure of the buyer. This is because any joint effort for maintaining or increasing the profitability is considered as investment in a true partnership.
- *Consider dedicated investments*: As soon as the Western buyer invests in the Chinese supplier, it perceives this as a sign of trust and long-term orientation. One concrete measure is to buy the necessary tools for the supplier. This reduces its cost. Even if it might be a relatively low investment and the effect on the supplier's cost is low, the symbolic value is high. It is a short-term measure that should be thought of as soon as the supplier starts to complain about a lack of profit.
- *Accept higher prices*: The case studies often showed situations in which the supplier would not any longer deliver products to the initially negotiated prices. These initially negotiated prices often have a relatively low margin, so when the supplier grows strong and becomes more powerful, it would not accept these low margins anymore. This development should be taken into account from the beginning, when budgeting the investment in the relationship.
- *If necessary, go into a transition stage*: When realizing that the supplier has other customers with lower requirements (which means easier profit for the supplier), investments in the relationship should be continued for a while, in order to gain time while finding and developing another supplier (e.g., buying tools for the supplier, paying higher prices).
- *Broaden the basis of the relationship*: When meeting with Chinese suppliers, it should not be focused only on the direct counterpart or the team leader, but on the whole team. For instance, at least an initial meeting where both teams meet should be organized, including the operational people. This also tackles another possible problem, a high fluctuation rate of key personnel at the supplier.

- *Utilize Guanxi*: Chinese companies should be considered as partners. It should be thought of how both can grow, even if the relative share of buying volume is small. The focus should be on establishing a long-term relationship and building up social capital. Then the Chinese supplier will keep the Western buyer as a customer, even with small relative buying volumes.

- *Develop a perspective for the partnership*: One way to provide a perspective for the partnership is an extension of the product range being bought from the Chinese supplier. This helps it grow not only quantitatively but also qualitatively. For instance, one measure several case companies would have liked to see, in order to develop the relationship toward a strategic partnership, is the joint discussion of new product areas for collaboration. Buyers often do not even ask how the supplier's competences improved over time due to other partnerships. Therefore, a continuous exchange of growth information in terms of market and technology would be an appropriate measure.

- *Improve your own processes*: When sourcing relatively low amounts of customized products with regular design changes, modularity and therefore more standardization in the product design should be considered, in order to decrease the costly design change issue. Then the Western buyer becomes more profitable for the supplier.

- *Stay honest*: When developing a second source, an *open information policy* should be followed. For instance, this means telling the supplier that it gets only 70% of buying volume and another supplier is being built up. This way the supplier has time to find a new customer, but can remain a second source for the original buyer. This is considered as a trustful relationship. Every Chinese would understand the situation because he or she would do the same (*smart* business behavior). This perception was identified as a major cultural difference when compared to Western business behavior.

9.8 Theoretical Implications and Limits of the Study

Conventional models for classifying power relationships are based on the dependence of the buyer or supplier in the relationship, mainly focusing on economic dependency in a stable growth environment. The described research broadens this perspective by providing a more comprehensive picture on the development of power relationships when sourcing in countries with higher growth rates than in Western countries and big cultural differences, by using the specific characteristics of the Chinese market. These aspects need to be integrated into existing models of buyer–supplier relationship management, which are mainly in cases based on pure economic factors.

The interviews were based on scenarios of Western European–Chinese relationships, although the interviewed suppliers also provided some examples of US and Japanese buyers. The number of case companies is rather small and covers only specific industries. Therefore, further research should cover relationships with non-European buyers to an appropriate extent, in order to foster the validity of the findings.

Additionally, the presented research was conducted on a qualitative level, which is appropriate for gaining knowledge in a practice-oriented and relatively new area. Further research should evaluate and validate the findings on a quantitative level.

References

Autry, C. W. and S. L. Golicic. 2010. Evaluating buyer–supplier relationship-performance spirals: A longitudinal study. *Journal of Operations Management*, 28:87–100.

Bacharach, S. B. and E. J. Lawler. 1981. *Power and Politics in Organizations: The Social Psychology of Conflict, Coalitions, and Bargaining*. Jossey-Bass, San Francisco, CA.

Bensaou, M. 1999. Portfolios of buyer-supplier relationships. *Sloan Management Review*, 40:35–44.

Caniels, M. C. J. and C. J. Gelderman. 2005. Purchasing strategies in the Kraljic matrix: A power and dependence perspective. *Journal of Purchasing and Supply Management*, 11:141–155.

Caniëls, M. C. J. and C. J. Gelderman. 2007. Power and interdependence in buyer–supplier relationships: A purchasing portfolio approach. *Industrial Marketing Management*, 36:219–229.

Fan, Y. 2002. Questioning Guanxi: Definition, classification and implications. *International Business Review*, 11(5):543–561. Available at http://bura.brunel.ac.uk/handle/2438/1279.

Kraljic, P. 1983. Purchasing must become supply management. *Harvard Business Review*, 61:109–117.

Nyaga, G. N., J. M. Whipple, and D. F. Lynch. 2010. Examining supply chain relationships: Do buyer and supplier perspectives on collaborative relationships differ? *Journal of Operations Management*, 28:101–114.

Oehmen, J., P. Gruber, M. von Bredow, and R. Alard. 2009a. Risk minimization in global sourcing by managing the bargaining power in buyer–supplier relationships. In: *Proceedings of Third International Conference on Changeable, Agile, Reconfigurable and Virtual Production (CARV)*. Munich, Germany.

Oehmen, J., P. Schönsleben, M. von Bredow, P. Gruber, and G. Reinhart. 2009b. Strategische Machtfaktoren in Kunden-Lieferanten-Verhältnissen. *Industrie Management*, 25:29–33.

Chapter 10

Outsourcing Design to Asia: ODM Practices

Qi Feng and Lauren Xiaoyuan Lu

Contents

10.1 Introduction

During the last few decades, the growing demand for electronics and computer products has given rise to a new breed of manufacturing firms in Asia, the so-called *original design manufacturers* (ODMs), for example, Flextronics, Quanta, Compal, etc. These companies provide design, manufacturing, and other value-added services to original equipment manufacturers (OEMs). In this chapter, we provide an overview of this new outsourcing practice by analyzing its key characteristics, decision factors, and implementation challenges.

Interestingly, the geographical regions that ODM firms are currently located in overlap significantly with the eastern part of the Silk Road, including China, South Korea, Singapore, Taiwan, etc. The historical routes for trading luxury goods between the East and the West are now part of the bigger global trading network for knowledge, technology, and skilled labor force. We hope that our survey on the latest ODM activities will provide insights for managers contemplating the decision to outsource product design to companies in these regions.

10.2 Growing Trend of Outsourcing Design to Asia

Outsourcing is widespread in diverse industries and for various activities. In this chapter, we focus our discussion on the electronics and computer industry, in which the ODM business model has received widespread popularity and also gained maturity since its inception.

10.2.1 Electronics Manufacturing Service Industry

The industry of electronics manufacturing service (EMS) started in 1981 when IBM decided to outsource the assembly of its computer products to contract manufacturers (CMs). This decision has since changed the landscape of electronics manufacturing. Three decades later, most electronics and computer OEMs outsource all or part of their manufacturing activities to CMs.

The growth of the EMS industry coincided with the spread of the Internet and the emergence of a wide range of consumer electronics products during the 1990s and 2000s. Worldwide revenues from the EMS industry reached USD 250 billion in 2009 and are forecasted to exceed USD 370 billion by 2014

(IDC, 2010). Figure 10.1 displays the estimated worldwide revenues of the EMS industry for the period 2005–2009 and the revenue projections for the period 2010–2014. Despite the recent global economic recession, it is believed that the EMS industry will bounce back from the downturn and continue to expand rapidly in the next decade with an estimated annual growth of 8% (IDC, 2010).

Outsourcing services provided by EMS companies were traditionally limited to manufacturing, that is, component production or final product assembly. In the 1990s, branded OEMs started offloading their manufacturing facilities and selling them to CMs in exchange for their provision of manufacturing services. OEMs' divesture of manufacturing facilities fueled the rapid growth of a handful of large CMs, including Foxconn, Flextronics, Jabil Circuit, Sanmina-SCI, and Celestica. Over the years, as these CMs became increasingly technology savvy, OEMs started to outsource product design and development related activities to these long-time manufacturing partners. This trend spawned a new outsourcing model—contract design and manufacturing (CDM)—in which an OEM hands over concept designs to a CDM firm and then both firms collaborate on physical design and development (Holloway and Hoyt, 2005). Although these outsourcing activities allow OEMs to reduce their own design content in the final products, they retain most of the control over the design decisions because they provide the concept/functional designs that dictate the physical/technical designs subsequently developed by the CDMs.

While large CMs expanded their business spectrum to provide CDM services for OEMs, another new breed of niche players, including Quanta, Compal, and Wistron, were also growing rapidly in providing design and manufacturing services. These companies, the so-called ODMs, followed a different path.

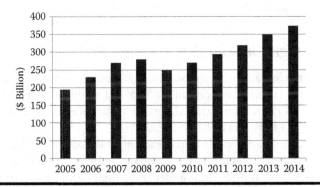

Figure 10.1   Estimated worldwide revenues of the EMS industry. (From IDC, EMS/ODM outlook, recovery tied to consumer electronics, 2010, available at http://www.ventureoutsource.com/contract-manufacturing/industry-pulse/ems-odm-outlook-recovery-tied-to-consumer-electronics)

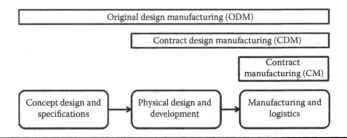

Figure 10.2 Outsourcing models and their associated phases in a product development cycle.

They designed and developed their own products and sold them "off-the-shelf" to OEMs (Huckman and Pisano, 2006). Figure 10.2 illustrates the three outsourcing models and their associated phases in a product development cycle. In the CM or the CDM model, OEMs own the intellectual property of designs and hold nearly complete control over the characteristics of a new product. The ODM model, in contrast, shifts both the control of product design and the associated intellectual property to the suppliers.

It is worth noting that the business press often does not distinguish between CDM and ODM activities explicitly. This is partly because the boundary between the two models is getting blurred and there exist many shades of gray among them in terms of the amount of design content owned by a supplier versus an OEM. Nowadays, it is common for EMS firms to provide a broad spectrum of service offerings ranging from CM and CDM to ODM. In the rest of the chapter, for convenience, we will use ODM to refer to all outsourcing activities consisting of both design and manufacturing components.

10.2.2 Design Outsourcing Activities in Asia

Design outsourcing activities are particularly concentrated in the Asia Pacific region, which is estimated to account for 92.5% of worldwide ODM revenue in 2010 (Wang, 2005). It is forecasted that the revenues of Asia Pacific ODMs will reach USD 110.4 billion in 2010 (Wang, 2005).

Taiwan, China, and South Korea lead in ODM activities. The history of Quanta Computer, a Taiwan-based company, mirrors the journey of many other Asian ODMs. In 1988, the founder of Quanta, Barry Lam, started designing his first notebook prototype and took it to a trade show, hoping to attract large computer OEMs (St. James Press 2002). The reaction to Lam's prototype was not immediate, but gradually orders came in. Quanta began production of its first notebook PC in 1990, and became the largest computer ODM in the 2000s, supplying 33% of the world's notebook PC production by 2007 (Kawakami, 2008).

Since the first ODM-made laptop was sold to the marketplace, the product category of design outsourcing has expanded widely over the years. According to Gartner Research, the top 10 product categories that drive the growth of design outsourcing activities in Asia are:

■ Mobile PCs (including laptops, netbooks, tablets, etc.)
■ PC motherboards
■ LCD monitors
■ Optical disk drives
■ PDAs, PDA phones, pocket PCs, and GPS PDAs
■ Servers
■ Digital audio players
■ Digital still cameras
■ Digital mobile handsets
■ LCD TVs

The spread of ODM activities to many product categories has some underlying drivers. ODMs are facing saturated markets and declining margins for their traditional product segments including notebooks, computers, and motherboards. They are also constantly pressured to lower cost by their OEM customers, who engage in fierce end-product competition. These challenging market conditions are compelling ODMs to keep innovating in their existing product categories and to expand to other segments (Wood, 2009).

10.3 Driving Factors of Design Outsourcing

Design outsourcing started to gain momentum in the late 1990s when the demand for computer, communication, and consumer electronic products experienced an explosive growth. In this section, we discuss the economic and managerial factors driving this growth.

10.3.1 ODMs Gain Design and Engineering Capabilities

Three decades ago, the EMS industry mainly consisted of CMs. These manufacturers, over the years, have not only accumulated extensive knowledge about manufacturing OEMs' products, but also gained expertise in managing complex engineering processes. Through technology acquisition and internal development, CMs have gradually built up technical know-how in product design and development. The ODM model emerged as engineers in the Asia Pacific region gained sufficient technological capability to design their own products from scratch, thanks to years of experience working with OEMs on the products they outsourced to the region.

The high concentration of ODM firms in Asia benefited from the colocation of electronics components supply chains and the growing body of skilled engineers in the region (Dedrick and Kraemer, 2006). In Taiwan, highly experienced engineers are available in electronic and mechanical design thanks to Taiwan's historical specialization in the PC industry, and notebooks in particular.

10.3.2 OEMs' Need to Streamline Product Development Processes

Nowadays OEMs are pressured to continuously bring new products to market with speed and quality. This is especially true in the electronics and computer industry, where product life cycles have shortened significantly—it is not uncommon for a consumer electronics OEM to introduce a new version of a product every 6 months. To stay ahead of the game, OEMs must be capable of efficiently managing their product development processes to meet the time-to-market and time-to-volume targets.

Large OEMs like Hewlett-Packard (HP) and Dell typically manage hundreds of products and several dozen product families. Their R&D resources can be spread thin. Purchasing off-the-shelf products built by ODMs becomes an attractive option for OEMs to expand their product portfolio. When OEMs fall behind in developing next generation products, buying off-the-shelf products from ODMs becomes a must to avoid losing market share to competitors (Carbone, 2003). Large ODMs have efficient production capabilities and well-managed subcontracting relationships with local suppliers. They have also formed close relationships with local component suppliers who provide technical support for ODMs' design and manufacturing processes. Because of these, ODMs are able to provide a volume production in a short time horizon for their OEM clients. Moreover, the ODMs' roles in engaging component suppliers cannot be easily replaced by OEMs—in many cases, the identity of OEMs is kept anonymous from suppliers (Pick, 2004).

Quanta was credited for salvaging the HP name in the notebook computer market (St. James Press, 2002). In 1999, the notebook division of HP was at the brink of being shut down when the company turned to Quanta for ODM services. According to a HP director in his interview with *Business Week* in 2001 (Einhorn, 2001), outsourcing to Quanta "saved our business" and represented "the biggest turnaround in HP's history."

10.3.3 OEMs' Strategic Shift to Focus on Core Competencies

Branded OEMs have made a strategic shift over the years to focus on their core competencies: branding, marketing, distribution, and innovation. Manufacturing activities were the first to be outsourced. In the late 1990s and early 2000s, the electronics industry experienced a wave of selling manufacturing facilities from OEMs to CMs. According to a Morgan Stanley report, the set of OEM assets that were candidates for divestiture could have been more than $10 billion per year in sales (Fleck and Craig, 2001). To focus their R&D resources on high-end and cutting-edge products, OEMs' outsourcing activities have gradually gone beyond manufacturing to include product design and development for low-end products. For example, Dell decided to outsource all of its netbook products, that is, low-end mobile PCs, to ODMs (Trainor, 2009).

OEMs' strategic shift affects not only their decision to outsource but also how their ODM strategy is implemented. The change of manufacturing strategy that Dell went through in the last few years illustrates an interesting example in this respect. Offering customized products to the mass market had been Dell's differentiating market strategy for many years. To support this business strategy, Dell conducted complex configure-to-order (CTO) operations and maintained three internal manufacturing facilities to support its major end markets. Traditionally, Dell had ODMs ship only the bare-bones mobile PCs without customizable parts to the three internal facilities, and then completed the final configuration according to customer requirements. Although Dell had enjoyed years of success in managing its supply chains efficiently, the declining margin in the PC market forced Dell to revamp its manufacturing strategy. Starting from 2007, Dell began to apply an outsourcing approach used by HP and Acer for years, which was to let ODMs ship completely built mobile PCs directly to the OEM's regional warehouses in the end markets (iSuppli, 2009). This standardization of manufacturing outsourcing strategy across major PC makers validates the ODM model as the right fit for the increasingly commoditized PC market.

10.3.4 Benefits of Concurrent Engineering and Design-for-Manufacturability

Because both product design and manufacturing are performed by ODMs, the ODM model offers opportunities for concurrent engineering and design-for-manufacturability. Collaboration between design and manufacturing, a common challenge in the traditional CM practices, is greatly facilitated. This potentially generates large cost savings for production and shortens time-to-market for OEMs' products. With concurrent engineering, design, development, manufacturing, and other functions are integrated to reduce the total time required to bring a new product to the market. The design-for-manufacturability approach leads to product designs that are easy and cost efficient to manufacture.

According to a government study (Holloway and Hoyt, 2005), 70% of a product's manufacturing cost is determined by the end of the design process. After that stage, only 35% of the cost reduction opportunities are left. Therefore, integration of design and manufacturing processes is crucial for making cost reductions. The ODM model offers this exact benefit, which is a major advantage compared to the traditional CM model or the CDM model. In the ODM model, a supplier takes the full responsibility of building the product, thereby facilitating efficiency and communications among design, engineering, and production staffs.

10.3.5 Serving New Market Entrants with No Prior Product Development Experiences

Easy access to design and development capabilities provided by ODMs has made entry to a product market feasible for firms that traditionally do not manufacture products. For example, mobile operators were able to offer their own-branded cell phones and smartphones thanks to the ODM model. In the wireless handset ODM market, HTC differentiates itself by offering mobile operators customized phones based on products designed by HTC. As T-Mobile's Senior Vice President Cole Brodman put it, HTC is "a company that embraces things we want" (Yoffie and Kim, 2009).

NexusOne, the smartphone offered by Google in January 2010, is another example. The growth potential in the smartphone market segment has attracted entries by technology giants like Google. By partnering with HTC via an ODM model, Google was able to make its presence in the market of 3G smartphones in a relatively short period of time.

Even though the sales of NexusOne did not meet the expectation due to issues pertaining to distribution strategy, Google gained significant experience from this product. In March 2011, the company launched its second smart phone, Nexus S built by Samsung.

10.3.6 Academic Views on Design Outsourcing

The widespread of outsourcing practices has generated diverse interests in academia. Some studies have investigated how the outsourcing decision is affected by capacity pooling (Plambeck and Taylor, 2005), allocation of demand risks (Ulku et al., 2007), and learning-by-doing (Anderson and Parker, 2002;

Gray et al., 2009). Other studies (e.g., McGuire and Staelin, 1983; Cachon and Harker, 2002; Gilbert et al., 2006; Arya et al., 2008) suggest the strategic benefit of outsourcing in mitigating competition in the end-product market. All these studies, while mainly focusing on manufacturing outsourcing, also help to understand the driving forces of design outsourcing.

There are only a few papers so far that specifically study design outsourcing. Parker and Anderson (2002) use case study data to illustrate how a large PC manufacturer acts as a supply chain integrator to coordinate product development after outsourcing the majority of its design and manufacturing activities to a network of suppliers. Cui et al. (2009) examine 24 outsourced development projects at Siemens and identify some common drivers of success such as project-specific partner competencies and maintaining in-house competence. Iyer et al. (2005) use a principal-agent framework to study a "core" buyer's product specification and production decisions in an outsourcing relationship.

As they take more responsibilities in the value chain of a product, ODMs gain in bargaining power. This, however, may not hurt the OEMs. By leaving a larger profit margin to ODMs, OEMs may effectively mitigate competition, as indicated by Feng and Lu (2010a). Their study also suggests that the increased bargaining power of an ODM in an outsourcing relationship may encourage design outsourcing by those OEMs who possess much weaker bargaining power than their competitors.

There is certainly a need for more research to deepen our understanding of design outsourcing practices. For example, how would the shift of design responsibility affect technology innovation? Would the knowledge transfer from OEMs to ODMs sustain a win–win outcome in the long term? Should OEMs always outsource designs for low-end products and retain those for high-end products in-house?

10.4 Issues and Challenges in Managing Design Outsourcing

Despite its growing popularity, concerns and reservations about the ODM business model have also emerged. While many managerial challenges exist for outsourcing in general, such as misaligned incentives, communication issues, cultural shocks, lack of information visibility, and risk of local regulatory environment, we focus on those issues specific to design outsourcing. In Sections 10.4.1 through 10.4.6, we discuss issues and challenges concerning practitioners. In Section 10.4.7, we summarize major insights from academic studies and suggest directions for future research.

10.4.1 Diluted Product Differentiation

Because the ODM market is dominated by a handful of large suppliers, it is inevitable that competing OEMs outsource to a common ODM. For example, Quanta Computer serves Apple, Dell, Gateway, HP, and Sony. OEMs increasingly worry about their outsourced work being leveraged to serve competitors. To ease the concerns, ODMs often setup separate divisions to trade with OEMs independently. Quanta Computer consists of several business units with independent R&D teams, sales managers, and factories to serve large OEM customers.

However, inter-divisional resource sharing may not be completely eliminated. Because the products designed by ODMs are often highly commoditized, there can be a tremendous benefit from scope economies in leveraging resources and knowledge across products of different OEMs. It is in fact not uncommon that similar products built by an ODM are marketed under different brands. Quanta Computer built Acer's TravelMate 800, Legend's Soleil A8202, and Sanyo's Skywalker 3100, which were derived from a common Centrino-based prototype. Samsung Electronics, who competed in the notebook ODM market in the early 2000s, designed Gateway 200 based on Dell Latitude X200 (Tzeng and Chang, 2003). These cost-reduction tactics have led to a proliferation of branded products with almost identical functionalities.

10.4.2 Fostering Competition by New Market Entrants

In the early days of design outsourcing, OEMs helped ODMs to build capabilities by transferring expertise in both management and engineering skills. While most ODMs focus on providing outsourcing services, OEMs' concern of ODMs becoming potential competitors is not unfounded. BenQ's aggressive pursuit of own-brand strategy led to its divorce with Motorola in 2004, which then accounted for two thirds of the former's revenue in the handset business. Such a transition could also lead to pitfalls to ODMs—after six consecutive quarters of loss, BenQ decided to refocus on ODM business in April 2005. Even for those who eventually built their own names, the journey can be long and painful. Acer Inc., now the fourth largest PC maker, made the transition from an anonymous ODM to a global brand. Yet Acer, stepping from its low brand recognition and lack of international expertise, had stumbled years on the way up.

As mentioned in Section 10.3.5, ODMs have enabled large retailers like Best-Buy and mobile operators like T-Mobile to sell their own-brand products, without acquiring in-house design capacities. This also forces OEMs to face potential competition from their downstream firms.

10.4.3 ODMs' Dependence on OEMs

Heavily relying on OEMs, ODMs are vulnerable to a sudden drop in business volume and may take long to recover (Deffree, 2009). For example, after the Internet bubble burst in 2001, Flextronics' cash earnings fell by more than half and only started to regain business in mid-2004.

ODMs also constantly face the risk of losing business to firms in emerging markets. One of the main benefits offered by ODMs is low-cost labor, which depends heavily on the local economy. ODMs migrate to other locations once such benefit is depleted. The geographic focus of the electronics supply network has shifted from Western countries to Japan in the 1960s, followed by a move to Korea and Taiwan in the 1980s and another one to China in the 2000s. Recently, the industry started to search for labor arbitrage in India and East Europe. Such migrations have fostered growing local players, who will soon change the dynamics of the industry.

10.4.4 Product Quality

Arguably, during the past decade, consumers are getting used to product recalls. The electronics industry is not alone in this wave. The US Consumer Product Safety Commission estimates an annual loss of over USD 700 billion for product-safety related incidents. In U.K., the number of product recalls rose from 112 in 2004 to 253 in 2007. The rising figures are partly due to the increasing awareness of potential hazards and new product safety legislation introduced. Firms' attempt to stay on top of competition also contributes to failures of premature product launches with uncovered defects. Due to fierce competition, firms are often pressured to push new products into market without being able to uncover all potential quality issues. Consumers and analysts sometimes put the blame on the suppliers—more outsourcing means more product recalls. Mattel's recall of Chinese-made toys with over-limit lead in 2007, Toyota's recall of vehicles with defective pedal in 2010, and McDonald's recall of Shrek Cup with dangerous cadmium in 2010 all raised concerns over outsourcing.

In the electronics industry, because the OEM–ODM relationship is more robust, a success requires efforts by both parties and a failure is hardly the fault of any single one. When the iPhone4's antenna problem was uncovered shortly after the product was released, consumers were disappointed with both Apple and Foxconn because "it does not take a radio-frequency engineer to know that stainless steel is conductive." The blurry boundary between ODMs and OEMs on the value chain could potentially dampen their quality incentives. Practitioners have long acknowledged that building strategic partnerships with ODMs is essential to the success of design outsourcing (Delattre et al., 2003).

10.4.5 Working Conditions in ODM Firms

In June 2010, Foxconn, the manufacturer of iPhone4, probably caught more attention than the product launch itself. After 14 suicides and suicide attempts, the company with 800,000 Chinese employees was accused for its military-style efficiency. The names of its partners, Apple, Dell, HP, and Lenovo, appeared along with Foxconn in media. Foxconn's sweatshop-like working conditions, such as long working hours, below-dollar hourly wages, and verbal abuse by superiors, have been known among electronics producers. However, labor issues have been prevalent across Asian companies for years. For example, 74% of the member companies of Electronic Industry Citizenship Coalition (EICC) urged the need for reducing long working hours. Moreover, exposure to toxic substances can also be a big concern. Just not long before Foxconn scandal, a young worker died of leukemia at Samsung's On-Yang facility in South Korea. By then, the company had documented 23 cases of workers suffering blood cancer, nine of whom died (Bormann and Plank, 2010). Such scandals can do a deep and long damage to the firms' reputations. While squeezing for cost savings, both OEMs and ODMs must realize their responsibilities toward improving working conditions.

10.4.6 Global Sustainability

Nowadays, brand owners see the growing need to proactively fulfill their social responsibility in the name of sustainable development. Consumers often perceive environmental friendliness as a positive sign for products. OEMs' largest impact on environment is through their products (e.g., 97% of Apple's carbon footprint is directly related to products). According to International Data Corporation (IDC), about one billion computers will potentially be scrapped in 2010.

To reduce their environmental impact, OEMs have long been marketing their brands with ENERGY STAR® and eco labels, and advocating their green programs. IBM initiated its Product Stewardship program in 1991, and HP established its design for environment program in 1992. Many have signed on to electronic industry code of conduct, which details strict guidelines for safe working conditions and environmentally responsible practices, provides tools to audit compliance, and helps companies audit progress.

Furthermore, OEMs usually provide their ODM suppliers with environmental guidelines and insist on carbon footprint reduction during product life cycle and beyond. The major considerations include reducing material use, adopting recycled or environmentally preferred materials, improving energy consumption and efficiency, allowing product upgradeability and recyclability at the end of life, and ensuring disposal safety. These considerations require both OEMs and

ODMs to be innovative in designing and developing eco-friendly product, while meeting the cost-saving objective.

This, however, is only one end of the equation. While OEMs are presenting their green products, ODMs, being located in regions with typically lower environmental standards, may be operating off consumers' radar. Thus, it is crucial for OEMs to conduct thorough supplier audits for complying with mutually agreed environmental standards. Otherwise, ODMs' environmental impact, for example, pollution, waste and emissions, may be hidden behind the OEMs' campaigns of their carbon footprint reduction.

10.4.7 Insights from Academic Studies on Outsourcing

The challenges and issues in outsourcing practices have received much attention from the academic community (e.g., Lee and Tang, 1996; Iyer et al. 2005). While most of the research papers focus on manufacturing outsourcing, they also shed light on design outsourcing.

Several studies indentify potential pitfalls for OEMs seeking low-cost outsourcing strategies. Transferring design and manufacturing to emerging market not only requires substantial initial investment by the OEMs (Galbraith, 1990), but also lowers the barrier for potential local imitators to enter the market and stirs competition. A study by Sun et al. (2010) suggests that less transfer to such emerging markets may be beneficial to OEMs when the potential cost saving is high. OEMs may not achieve their cost-saving goals even in the absence of potential market entrants. A supplier's bargaining position strengthens when his cost becomes lower. As Feng and Lu (2010b) point out, this may backfire on OEMs who outsource to low-cost suppliers. How to get around the trap? Feng and Lu suggest that allowing a certain degree of incentive misalignment between negotiating parties (e.g., via a wholesale-price contract), as opposed to fully coordinating their decisions (e.g., via a two-part tariff), can lead to a win–win outcome.

Specifically analyzing the ODM business model, Feng and Lu (2010a) confirm practitioner's concern that scope economies derived from making products for multiple OEM clients can induce ODMs to significantly reduce product differentiation. They also suggest that such an incentive can be mitigated by letting ODMs acquire a high profit margin in contract negotiation.

The academic research on design outsourcing is, as mentioned earlier, underdeveloped. More work is needed to analyze the challenges and suggest managerial guidance for the practice. First, OEMs and ODMs need to take different measures to address environmental considerations in product designs. How to align their incentives to achieve a green value chain is an important question yet to be answered. This would require a general framework that incorporates the design network in the reversed logistics network. While there is a vast literature on managing the latter (e.g., Dowlatshahi, 2003; Savaskan et al., 2004; Debo et al., 2005; Ferrer and Swaminathan, 2006), its linkage to design outsourcing strategies has not been carefully examined. Second, successes of high quality products require joint effort from OEMs and ODMs, who become more and more dependent upon each other. It is, however, not clear how the concern of brand reputation affects outsourcing strategies and design decisions. For example, should competing OEMs outsource to the same ODM and share the same key components in their products? Without much differentiation, if one's product fails, others would most likely be in the same boat. Addressing these issues requires a reexamination of the supply chain literature on component offering (e.g., Chernev, 2003; Bernstein et al., 2011).

10.5 Conclusions

Design outsourcing to ODMs is a rapidly growing practice but has received little attention in the academic literature. We intend to use this overview as a call for more research work on this relatively new business model, which is still going through changes as the business environment for outsourcing evolves. We believe that investigating the managerial issues of design outsourcing may generate fruitful findings that are valuable to both academics and practitioners.

References

Anderson, E. and G. Parker. 2002. The effect of learning on the make/buy decision. *Production and Operations Management*, 11(3):313–339.

Arya, A., B. Mittendorf, and D. E. Sappington. 2008. The make-or-buy decision in the presence of a rival: Strategic outsourcing to a common supplier. *Management Science*, 54(10):1747–1758.

Bernstein, F., G. Kök, and L. Xie. 2011. The role of component commonality in product assortment decisions. *Manufacturing and Service Operations Management*, 13(2):261–270.

Bormann, S. and L. Plank. 2010. Working conditions and economic development in ICT production in central and Eastern Europe. *World Economy, Ecology and Development*, Berlin, Germany.

Cachon, G. and P. T. Harker. 2002. Competition and outsourcing with scale economies. *Management Science*, 48(10):1314–1333.

Carbone, J. 2003. ODMs offer design expertise, quicker time to market. *Purchasing*, 132(1):11–14.

Chernev, A. 2003. When more is less and less is more: The role of ideal point availability and assortment in consumer choice. *The Journal of Consumer Research,* 30(2):170–183.

Cui, Z., C. H. Loch, B. Grossmann, and R. He. 2009. Outsourcing innovation: Comparison of external technology providers to Siemens uncovers five drivers of innovation success. *Research Technology Management,* November–December, pp. 54–63.

Debo, L. G., B. L. Toktay, and L. V. van Wassenhove. 2005. Market segmentation and product technology selection for remanufacturable products. *Management Science,* 51(8):1193–1205.

Dedrick, J. and K. L. Kraemer. July 2006. Is production pulling knowledge work to China? A study of the notebook PC industry. *Computer,* 39(7):36–42.

Deffree, S. 2009. Recession slowed growth of design outsourcing in 2008. Electronic Design, Strategy, News (EDN) August 17, 2009.

Delattre, A., T. Hess, and K. Chieh. 2003. Strategic outsourcing: Electronics manufacturing transformation in changing business climates. *Research Report,* Accenture.

Dowlatshahi, S. 2003. Developing a theory of reverse logistics. *Interfaces,* 30(3):143–155.

Einhorn, B. 2001. Quanta's quantum leap: The Taiwan notebook maker's business could surge by 50%. *Business Week,* Linkou, Taiwan, November 5, 2001.

Feng, Q. and L. X. Lu. 2010a. Design outsourcing in a differentiated product market: The role of bargaining and scope economies. Working Paper.

Feng, Q. and L. X. Lu. 2010b. Is outsourcing a win–win game? The effects of competition, contractual form, and merger. Working Paper.

Ferrer, G. and J. M. Swaminathan. 2006. Managing new and remanufactured products. *Management Science,* 52(1):15–26.

Fleck, S. A. and S. D. Craig. 2001. *Solectron.* Morgan Stanley Dean Witter, New York, June 27.

Galbraith, C. S. 1990. Transferring core manufacturing technologies in high-technology firms. *California Management Review,* 32(4):56–70.

Gilbert, S., M. Y. Xia, and G. Yu. 2006. Strategic outsourcing for competing OEMs that face cost reduction opportunities. *IIE Transactions,* 38:903–915.

Gray, J. V., B. Tomlin, and A. V. Roth. 2009. Outsourcing to a powerful contract manufacturer: The effect of learning-by-doing. *Productions and Operations Management,* 18(5):487–505.

Holloway, C. and D. Hoyt. 2005. Flextronics: A focus on design leads to India. Stanford Graduate School of Business, Stanford, CA, Case Study.

Huckman, R. S. and G. P. Pisano. 2006. Flextronics International, Ltd. Harvard Business School, Boston, MA, Case Study.

IDC. 2010. EMS/ODM outlook, recovery tied to consumer electronics. Available at http://www.ventureoutsource.com/contract-manufacturing/industry-pulse/ems-odm-outlook-recovery-tied-to-consumer-electronics (accessed: October 31, 2010).

iSuppli. 2009. Dell strategy: Mobile PC outsourcing benefits and drawbacks. Available at http://www.ventureoutsource.com/contract-manufacturing/industry-pulse/2009/dell-strategy-mobile-pc-outsourcing-benefits-and-drawbacks

Iyer, A. V., L. B. Schwarz, and S. A. Zenios. 2005. A principal-agent model for product specification and production. *Management Science,* 51(1):106–119.

Kawakami, M. 2008. Exploiting the modularity of value chains: Inter-firm dynamics of the Taiwanese notebook PC industry. Institute of Developing Economies Discussion Paper Series No.146.

Lee, H. and C. Tang. 1996. Managing supply chains with contract manufacturing. *Asian Journal of Business and Information Systems*, 1(1):11–22.

McGuire, T. W. and R. Staelin. 1983. An industry equilibrium analysis of downstream vertical integration. *Marketing Science*, 2(2):161–191.

Parker, G. G. and E. G. Anderson. 2002. From buyer to integrator: The transformation of the supply chain manager in the vertically disintegrating firm. *Production and Operations Management*, 11(1):75–91.

Pick, A. 2004. Working with ODMs—A Component supplier's perspective. Available at http://www.emsnow.com/newsarchives/archivedetails.cfm?ID=4813 (accessed: October 31, 2010).

Plambeck, E. L. and T. A. Taylor. 2005. Sell the plant? The impact of contract manufacturing on innovation, capacity, and profitability. *Management Science*, 51(1):133–150.

Savaskan, R. C., S. Bhattacharya, and L. N. van Wassenhove. 2004. Closed-loop supply chain models with product remanufacturing. *Management Science*, 50(2):239–252.

St. James Press. 2002. Quanta Computer Inc. *International Directory of Company Histories*, Vol. 47.

Sun, J., L. Debo, S. Kekre, and J. Xie. 2010. Component-based technology transfer in the presence of potential imitators. *Management Science*, 56(3):536–552.

Trainor, S. 2009. Private communication. Dell Corporate, Roundrock, TX.

Tzeng, D. and S. Chang. 2003. ODM notebook strategy: Reuse the same design for as many customers as possible. *Digitimes*, March 17.

Ulku, S., B. L. Toktay, and E. Yucesan. 2007. Risk ownership in contract manufacturing. *Manufacturing and Service Operations Management*, 9(3):225–241.

Wang, J. 2005. Key electronics production and semiconductor use by Asia/Pacific ODMs. Gartner Research Dataquest, Stamford, CT.

Wood, L. 2009. Research and markets: The Asia Pacific original design manufacturing ODMs to expand into other commoditized markets to offset economic slowdown and rising costs. Available at http://www.reuters.com/article/idUS162298+22-May-2009+BW20090522 (accessed: October 31, 2010).

Yoffie, D. and R. Kim. 2009. HTC Corp. in 2009. Harvard Business School, Boston, MA, Case Study.

Chapter 11

Milk Collection at Nestle Pakistan Ltd.

Arif Iqbal Rana and Mohammad Kamran Mumtaz

Contents

11.1 Introduction

Pakistan is on the crossroads of the silk route. This is where the route from China to Iran intersects with that from India to Central Asia. It is not surprising that when the highway linking Pakistan to China (through the Karakoram Range in the Himalayas) was built, it was commonly called the "silk highway."

Pakistan is the fourth largest milk producing country in the world, with an annual production of 34 billion liters. Out of this, 27 billion liters are available for human consumption. Despite being the fourth largest producer of milk in the world, Pakistan cannot fulfill its demands. Average annual deficits ran at 1 billion liters in 2009–2010. Nevertheless, dairy plays an important role in Pakistan's economy as it contributes 14% to the country's overall Gross Domestic Product (GDP). The importance of dairy for Pakistan may also be judged by the fact that more than 20% of its entire population is directly or indirectly involved in the dairy business.

Pakistan has come a long way in milk production in just one generation. From the late 1970s when selling "milk" was looked down upon socially, to 2009, when many in the country are predicting a "white revolution." The country is already the fourth largest milk producer in the world despite the fact that almost all milk farms are in the informal sector. However, at least a couple of dozen large dairy farm projects are underway. Almost all the major industrial groups in the country either have a dairy farm project underway or are considering starting one. However, milk collection in the country still has many major issues: from serious quality problems with collected milk, to major drop in buffalo milk production in the summer, to still dominant social customs like not selling milk on the 11th of the moon (11th of the moon is an important date in Sufi tradition in the Indo-Pak subcontinent; hence many herd owners donate their milk to charity instead of selling it on this date).

11.2 Pakistan and Milk

Pakistan has not been able to take full advantage of its dairy strength which is evident from the performance of the country's policy initiatives in the dairy sector. The first five year plan (1955–1960) introduced a scheme whereby the government would buy milk (after testing it for quality) from gawalas (milkmen) residing in gujar (a cast of milkmen) colonies, pasteurize it, and supply to public in sealed bottles from registered milk depots. The plan suggested that most of the milk would be produced in villages near major cities by milkmen who would specialize in this business. It also suggested that these milkmen would organize themselves into cooperatives for assembling, transporting, and processing of milk. The plan though well articulated faced problems in implementation. The second (1960–1965) and third (1965–1970) five year plans laid little emphasis if any on the development of dairy sector. However, work on the agenda in the first five year plan continued. As a result the first milk supply scheme as envisaged in the plan became operational in Karachi in 1965 and in Lahore in 1967. Subsidized milk was supplied to low income families with support from the United Nations Children's Fund (UNICEF). However, these supply schemes had to be abandoned owing to financial losses and little support from successive governments later on.

The milk processing industry in Pakistan flourished in the 1970s as a result of huge investment in it by the private sector, which established 23 milk processing plants around three big cities, i.e., Karachi, Lahore, and the twin cities of Rawalpindi and Islamabad. These plants did not turn out to be successful because of short shelf life of milk and little acceptance of processed milk by consumers.

The government gave incentives for investment in the dairy sector in the late 1970s and early 1980s. Furthermore, investment in Milkpak (ultra-high temperature [UHT] treating milk plant by Packages Ltd., a part of the Ali Group) in the early 1970s and its success later in the century attracted several private players in the dairy sector. This investment was also supported by the introduction of aseptic packaging material for UHT treated milk by Tetra Pak Pakistan Limited. Several plants were set up in the 1980s. They were not successful though because demand for processed milk in the country was still low and many players could not sustain production costs at low levels of production. Later government's policies laid emphasis on establishing large-scale private enterprises while limiting its role to conducting research in dairy and providing conducive environment for growth of the sector.

11.3 Structure of Pakistan's Dairy Sector

Pakistan's dairy sector is characterized by fragmented small holder dairy farmers and unorganized farms mostly operating on noncommercial basis. Majority of the farms in Pakistan operate with small herd sizes, with over 82% of the farms owning less than six animals. These small farms are operated at subsistence or near-subsistence level, i.e., dairy is not a major source of income for them. A key feature of subsistence dairying is that women are involved in dairy while men concentrate on other tasks considered more important. Majority of the small holders keep a large portion (almost 75%) of the milk produced for home use. They use the milk produced to enhance their agriculture income. Primary purpose of animals is agriculture.

Furthermore, the small farmers do not have sufficient land to cultivate their own fodder. They cannot also afford costly ration for their animals. Hence more than half of the feeding requirement for these animals is provided by grazing. Small herd sizes also do not provide economies of scale and cost to retain them is usually higher than the revenue generated from selling of milk provided by these animals. Hence cost per liter is usually higher for small holders.

Over the last decade, many large farm projects have been initiated on a commercial basis by investors interested in reaping the benefits of the "white revolution." Large commercial farms usually with 30 or more animals are built in accordance with modern farm designs, i.e., they are built keeping in mind the requirements of better rearing, higher yields and long survival of the animals. For instance, these farms would usually have a bathing area for animals, drinking area with plenty of water supply, feeding area, and sheds to protect the animals from heat. The animals at these farms are usually fed on foods with high nutritional values. The feeds on these farms include concentrates, wheat, rice straw, berseem, sorghum, and maize (different types of animal feed).

Commercial farm owners have better knowledge of good farm management practices as compared to their counterparts. They employ modern dairy technologies to improve milking of animals. They also have access to better medical facilities for their animals. They take advantage of good breeding practices such as artificial insemination to improve the quality of animals and hence their yield. Better animal stock with good farm management practices and high-nutrition feed results in higher yield for commercial farms. However, because of low number of such farms (since these farms require hefty investments and high operating costs majority of the low income farmers cannot afford building and maintaining such farms) their contribution in overall milk production in the country is low.

11.3.1 Milk Supply Chain

Dairy supply chain in Pakistan may be classified into two major categories. The first is the formal sector supply chain and the second is the informal sector supply chain. The formal sector makes only 3% of the total milk produced and consumed in Pakistan. Rest of the 97% share of the milk market is taken by the informal sector. In the formal sector Nestle had the largest market share, i.e., 60% till 2008. This is before entry of Engro Foods into UHT treated milk business. Later Engro took larger part of the formal milk market but Nestle maintained its lead in terms of market share.

In the formal sector milk is collected by milk collecting agencies or processing plants that store the milk in cooling towers. This milk is then taken to the processing plants where the milk is processed and packed. Packaged milk is then taken to distribution centers from where it is distributed to retailers who then finally sell the milk to urban consumers.

Informal sector can further be divided into two categories. In one case farm milk is collected by gawalas (milkmen) who then sell the milk directly to consumers or the milk collectors. In the second case milk from farmers is directly collected by milk collectors who then sell the milk to retailers. These retailers then sell the milk to the consumers.

11.4 Characteristics of Milk Production Systems in Pakistan

Milk consumed in Pakistan comes from two major sources, i.e., buffaloes and cows. Buffalo milk constitutes 68% by volume while cow milk makes almost 32%. Buffalo milk is high in demand and therefore is sold at higher price as compared to cow milk (higher fat content in buffalo milk also makes it attractive for milk processing plants). The milk production and collection system existing in the country provides several challenges for players in the dairy business. Following is a description of some of the problems in which Pakistan's dairy sector exists.

The dairy sector in Pakistan is fragmented with large number of small players scattered over the country, though majority of milk production comes from the province of Punjab. The market is unregulated. This means that there is little government intervention in the transactions that take place between various stakeholders in the dairy business. The farm level prices are determined by market forces. Middlemen dispersed across the country exploit the poor farmers giving them low prices for their milk often delaying the payments. Large contractors often blackmail small-scale producers as well as processors.

Milk production in Pakistan is met by a number of other constraints as well. Firstly, milk yield per dairy animal in Pakistan is low. An animal of Pakistani breed gives between 1300 and 2400 kg of milk per annum as compared to yields of 6000 kg/animal/year in the rest of the world. A major cause of low yield is the stock of dairy animals in Pakistan which is not capable of producing large quantities of milk as compared to breeds in the west. Small-scale farmers, especially those in rural areas are not aware of and accustomed to practices of artificial insemination or crossbreeding which provides the opportunity of producing animals with potential for higher yield. Furthermore these farmers are not trained in modern dairy practices. These include methods for rearing and milking the animals. In a number of cases wrong milking methods lead to mastitis, a disease prevalent in cows and buffaloes. In addition to lack of training in proper animal husbandry practices, the farmers do not have sufficient knowledge of different types of feeds, their advantages, and precautionary measures to prevent animals from diseases. Furthermore, the low level farmers have little access to facilities like veterinary doctors and medicines. These problems not only reduce the milk yield per animal but also restrain animals' ability to be productive for longer period of time.

Second issue facing the milk producers is that milk productivity is seasonal. During summers, grasses and herbs dry out fast and as a result animals that graze on these grasses do not find enough food to eat. Furthermore, because of the heat the digestive system of animals becomes weak. As a result the yield decreases during summers. Milk production falls to 55% of its peak production level in mid-June (peak summer time). To aggravate the problem, the demand for milk and milk products increases during the summer season by 60% as compared to its value in December when there is ample supply of milk. This significant difference between supply and demand during lean season when the supply is short and flush season when the supply is ample and demand is low is a major cause of concern for operators in the dairy sector.

During lean season, a lot of farmers and suppliers prefer to sell to small, unorganized milkmen who can offer them higher prices for their milk. For example, in 2008 the milkmen could offer up to 5 rupees extra per liter to farmers as compared to the formal sector. These milkmen usually collect milk from around the major cities like Lahore, Karachi, Faisalabad, etc.; areas adjacent to these cities are strongholds of milkmen and no formal sector milk company has its milk collection setup in these areas. However, in summer months, these milkmen set up camps in remoter areas to collect milk because their core milk shed, i.e., areas around big cities, would be unable to fulfill demand. The milkmen can afford to pay a higher price for the farmers' milk because they do not process or package milk and thus no additional costs are incurred besides transportation. Additionally, a majority of the milkmen adulterate milk, adding water, fats, and other alien, often harmful, substances to the milk. If formal sector processors were to offer the same rates as them, their milk business would cease to be profitable, given the costs of processing, packaging, overheads, and quality.

A majority of milkmen adulterate milk which is an issue difficult to control. Sometimes farmers may themselves add water and other products to milk to increase its quantity. Hence ensuring quality of milk is a time consuming and costly activity for milk processors.

To add to the problems, the infrastructure in Pakistan makes milk handling and transportation difficult and costly. Lack of proper milk chilling systems (in Pakistan mostly ice is used to chill the milk; this method of chilling milk is not optimal and affects milk quality. Though small refrigeration are available, they are not portable, require electric power and are expensive) results in significant milk wastage in the supply chain as raw milk from the farm needs to be calm-chilled within 4 h of milking or otherwise it starts getting bad. According to World Bank reports, 15% of milk sold at farm gate gets lost before it reaches the milkmen while additional 5% gets wasted during transportation.

Finally, small-scale farming is itself a problem in a sense. The cost for feed, which is a major cost component in dairy farming, is high for small-scale farmers who cannot take advantage of economies of scale. These farmers do not have access to formal market channels and are prone to exploitation of middlemen who offer lower prices to these farmers for their products while they themselves retain huge margins from their own customers.

11.4.1 Nestlé Pakistan Limited

Nestlé Pakistan's roots could be traced to Milkpak Limited. Milkpak was part of a family group of businesses—the Ali Group—that spanned a number of activities, including razor blade and textile manufacture, insurance, vegetable oil, soap, and the management of Ford's auto assembly plant until its nationalization in 1973. The Ali Group was considered as one of Pakistan's leading industrial families. Milkpak was founded in 1979 to create a market for packaging materials produced by Packages Limited, a leading company in the Ali Group.

Packages Ltd. was established in 1956, in collaboration with AB Akerlund & Rausing of Sweden, to convert paper and board into packaging. Packages later integrated backwards into pulp and paper manufacturing. Packages had purchased a Tetra Laminator machine from Tetra Pak of Sweden, a company affiliated with Akerlund & Rausing, in 1967. This machine was designed for making packaging material for long-life UHT milk. When milk is heated at 130°C–150°C for 2 or 3 s, it can be stored without refrigeration for up to 3 months, packaged in Tetra Pak containers, which is very practical in countries like Pakistan. This machine was used very infrequently and in 1976 Packages decided to go into the UHT milk business to use it at its full capacity.

Milkpak Limited was incorporated in 1979 after Packages decided to invest in a 150,000 L/day UHT milk plant, at a cost of PKR. 90 million. Production started in 1981 and by 1987, Milkpak's product line had expanded from UHT milk to include fruit drinks and other dairy products, though UHT milk still accounted for an estimated 85% of company sales. Additional products included butter, UHT cream, and cooking oil.

Nestle Milkpak Limited was established in 1988 as a joint venture between Nestle S.A. and Milkpak Limited. In 1992, Nestle S.A. took over the management of Nestle Milkpak Limited. In 2004, the company was renamed Nestle Pakistan Limited. By 2005, Nestle Pakistan was in the business of manufacturing and marketing a wide range of brands of both dairy and nondairy products. Figure 11.1 illustrates the history of brand launches at Nestle Pakistan.

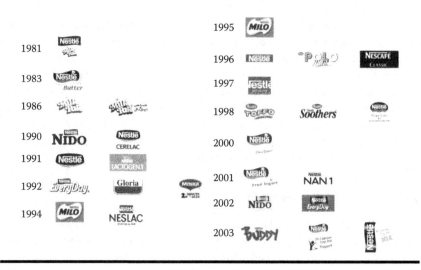

Figure 11.1   Nestle Pakistan Limited: History of product launches.

The corporate office of the company was located in Lahore at 308 Upper Mall. Nestle Pakistan Limited owned two factories. One was situated at Sheikhupura, referred to as the "Sheikhupura factory" (SKP), and the other at Kabirwala, district Khanewal, known as the "Kabirwala factory" (KBF). Nestle Pakistan acquired three water companies in Pakistan in 2001. These companies manufactured and retailed the brands AVA and Fontalia, which included sales of 5 gal home/office use bottle. Besides SKP and KBF, Nestle had three exclusive water factories, one in Islamabad manufacturing Nestle Pure Life and two in Karachi, manufacturing AVA and Fontalia. Nestle Pakistan's geographical presence is shown in Figure 11.2. The company's sales evolution is given in Figure 11.3.

11.4.2 Milk Collection at Nestle Pakistan

The core raw material of Nestle Pakistan was milk. Over the last 12 years, the company's prime concern had been to improve the quality and volume of milk for UHT processing and for other milk based products. Driven by its commitment to quality and having realized that only self-collection could eliminate its dependence on poor quality milk available from outside sources, the company successfully established its own collection system and expanded its operations over a very large milk shed area in Punjab. Owing to this tremendous growth in the volume of high quality raw milk, Nestle could boast of the largest milk collection network in the country, unmatched in size, productivity, and efficiency.

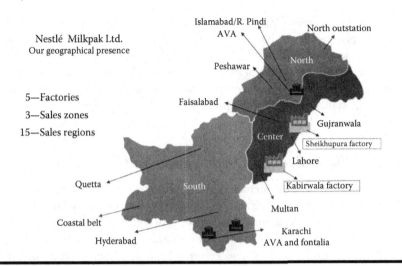

Figure 11.2   Nestle Pakistan Limited: Geographical presence.

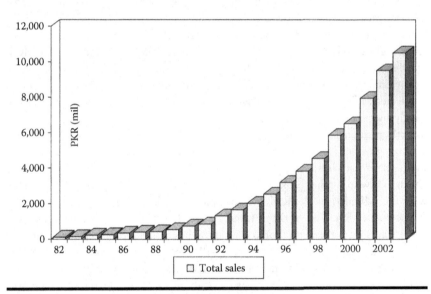

Figure 11.3   Nestle Pakistan Limited: Sales evolution.

Milk was collected through a vast network of village milk centers (VMCs), subcenters, and centers. At these centers, chillers had been installed to lower milk temperature to 4°C for preventing bacteria development during long hauls to the factories, which were undertaken by a large fleet of specially insulated tankers. In terms of quality, the milk collected by Nestle had to be low in sodium, high in fat and solid-non-fats (SNF), and very low in total plate count (TPC), which stated simply, means the bacteria count. As a service to farmers, Nestle established an extension service, staffed by qualified veterinary doctors, to assist them in vaccination and treatment of livestock, improved breeding, good animal husbandry practices, provision of high yield fodder seed, etc. By taking professional help and guidance to their doorsteps, Nestle tried to create a mutually beneficial relationship with the farmers.

11.4.3 Milk Collection Organization and Processes

Milk consisted of mostly water and a small percentage of solids. These solids can further be subdivided into fats and SNF. Buffalos were more popular in Pakistan than cows when it came to milk. Nestle Pakistan collected both cow and buffalo milk. Buffalo milk was 85% water and 15% solids, out of which 6% was fat and 9% SNF. Cow milk was 87.5% water and 12.5% solids, with fats at 4.5% and SNF at 8%.

The milk shed was divided into 24 regions spanning most of southern Punjab, each headed by a regional milk collection manager (RMCM). The milk collection center in each region coordinated all activities within the region. Each region collected milk through 60–70 milk collection subcenters, VMCs, and feeding points. Territory incharges (TIs) were responsible for milk collection from up to 10 collection subcenters and feeding points in their assigned territories and reported to the RMCM. The role of the TI was to maintain a steady supply line by building relationships with suppliers and farmers. He was directly responsible for the development of milk collection in his territory and achieving targets. Volume targets for milk collection were developed by the top management at Nestle and passed on to the National Milk Collection Manager (NMCM) who further broke them down and passed on to the RMCMs. Targets were then divided by the RMCMs and passed on to the territory level with the TIs' consultation. The territory targets were then further divided among subcenters and were then monitored on a daily basis. TIs monitored best practices for hygiene, equipment maintenance, quality testing, and record keeping at the subcenter. They also gave some veterinary services to the suppliers. The milk collection organization at Nestle consisted of about 2200 people, out of which 400 were permanent company employees and 1800 were on contract.

Milk collection subcenters were housed in shops rented by the company in or near villages. One company employee, the collection agent, ran each subcenter and was responsible for the actual milk collection, quality testing, and maintenance of the facility and equipment. Standard equipment at a subcenter consisted of quality testing equipment and chemicals, a chiller, and a diesel generator. Nestle accounted for more than 80% of the total chillers present in the milk collection universe in Pakistan, with most of the competition using ice to chill milk. Nestle had been investing more than Pakistan Rupee (PKR) 100 million per annum on chillers since 1992. Feeding points collected milk from those locations whose milk volume did not justify investment in a subcenter setup. Feeding points collected milk in large drums and delivered to the nearest subcenter to be chilled. However, feeding point agents were not under direct control of Nestle.

Subcenters were Nestle's representation in the local village or town to collect milk from milk farmers as well as independent suppliers ("dodhis"). When milk was brought to the subcenter, the first task the collection agent performed was measuring the volume of the milk. Then he performed a series of quality tests. The first was the organoleptic test, i.e., the color, smell, and taste of the milk. The agent checked the milk against these three attributes for anomalies. If the milk passed the organoleptic test, the agent then measured the milk's specific gravity using a lactometer and its fat content with an alcohol precipitation test (APT). If these two tests were successful, he poured the milk into the chiller. The agent then issued an invoice to the farmer/supplier against which he/she could claim for any anomalies in payment. The company registered farmers/suppliers at the subcenter level. These registered farmers opened their accounts in the nearest bank and payment for their milk was credited to their accounts on a weekly basis. The registered farmers received incentives for registering like milk collection promotions, milk-boosting fodder and fodder additives, and free veterinary advice and treatment for their cattle. Each subcenter had between 30 and 40 registered farmers. Unregistered farmers were paid in cash up front every Wednesday. Cash for this payment was remitted to the account of the collection agent on a weekly basis, who withdrew it from the bank and kept it with him at the subcenter. These payments were usually very small, for farmers supplying milk infrequently and in small volumes. Payment for suppliers was calculated based on milk volume and quality using the formula

$$P = \frac{V \times (\%\text{fat} \times R)}{S}$$

where
  P is the payment due
  V is the volume supplied
  R is the standard rate of milk
  S is the standard fat content (6% for buffalo milk, 4.5% for cow milk)

R varied among subcenters depending on level of competition and milk supply around the year. R was highest in high-competition locations in summers, hovering between PKR. 17 and 19, whereas in flush season it could be as low as 14.

Milk collection took place twice a day. Major volume is collected in the morning, from sunrise till around 11 a.m. Cattle were milked usually around sunrise, after which farmers stored milk in their utensils with no chilling facilities. For this reason, subcenters had a collection cutoff time between 10 and 11 a.m., because milk stored at the farmers' premises for any longer was sure to have deteriorated in quality. The agent stopped collection at the cutoff time and continued to chill the collected milk, maintaining a temperature of 0°C–4°C. The same process was repeated for evening milk collection, starting at sunset.

Each subcenter had a time assigned at which the milk truck picked up the milk and delivered to the Collection Center. The milk trucks had a capacity of 8 or 10 ton. Milk was transferred from the chiller to the truck's holding area via a pump. When it was filled to capacity, it came to the collection center and deposited the milk in the collection center's chillers. In flush season, a milk truck often delivered milk to the collection center two to three times per collection, i.e., morning/evening. When milk was delivered to the collection center, all earlier quality tests were repeated on the milk, as well as additional tests to check for adulteration. If the milk passed all these tests, it was deposited in the chiller. If the milk quality was slightly out-of-norm, it was returned and sold to third parties. If the quality was grossly compromised, the milk was drained. The milk trucks were then thoroughly washed to remove any traces of milk that may compromise the quality of the next batch. Standard procedures for this cleaning process were followed and the practice was called cleaning in-place (CIP).

The day's collection was chilled to 0°C at the collection center. Near midnight, this milk was transferred to large insulated tankers of 20 and 25 ton capacity and shipped to the factory. The insulation and night weather helped in maintaining the milk's temperature below 4°C during the journey. At the factory, all previous quality tests were repeated and further tests were done. If the milk passed all tests, it was accepted. Otherwise it was returned to the collection center. A complete list of all these tests is given in Figure 11.4.

11.5 Management Issues in Milk Collection

Milk collection staff at Nestle in 2004 was low paid, demotivated, and suffered from low self-esteem. The milk collection function had always been looked down upon by the rest of the company. For Nestle Pakistan, 80% of whose business came from dairy, milk collection was the life and blood. Collection was very similar to the sales function, with people having similar responsibilities and work hours. However, people working in milk collection were paid less and were not given same respect as people in other functional areas of business. Until 2005, TI was called Milk Collection Supervisor (MCS). There was further

- Organoleptic test
- Alcohol precipitation test (APT)
- Clot-on-boiling test (COB)
- Lactometer reading
- Temperature measurement
- Starch test
- Glucose test
- Sodium measurement test
- pH measurement
- Acid test
- Refractive index measurement
- Sugar test
- Carbonate/bicarbonate test
- Urea test
- Ammonium sulfate test
- Milk powder addition test

Figure 11.4   Nestle Pakistan Limited: Complete list of quality tests performed on milk.

division within this title. Seventy percent of MCSs were on contract, despite many of them being attached to the company for up to 10 years. These contractual employees were called milk service providers (MSP). The division of MCSs and MSPs resulted in discrimination and friction between the two. In 2005 Nestle made the MSPs permanent employees of the company, merged the MCS and MSP positions into one, i.e., TI and brought their salaries to parity with the sales organization. This resulted in up to 300% increase in salary of some TIs and up to 100% for Subcenter Agents. At the same time the company fired a number of underperforming employees. In one instance the company fired eight people in a single day.

Until 2005, TIs were expected to arrange their transportation themselves. The chief means of transportation for them was motorbikes and many TIs had bikes over 10 years old. In 2005, the company invested PKR 23 million in new motorbikes and vehicles for the milk collection staff. Nestle furthermore appointed two dedicated training managers to help improve the performance of the subcenter agents. The training managers conducted a "pre" survey of the milk collection practices and a "post" survey after 8 months of training of all subcenter agents. The results are shown in Figure 11.5.

To motivate the milk collection staff, Nestle introduced incentive schemes both for its staff and suppliers based on the sales promotion model. For the staff, the company started a performance incentive plan whereby the staff received double their salary if they achieved 110% of the targeted volume for a month.

For the farmers, Nestle started a Loyalty Incentive scheme. As the lean period for milk starts in May, Nestle kicked off its Loyalty Scheme on April 1.

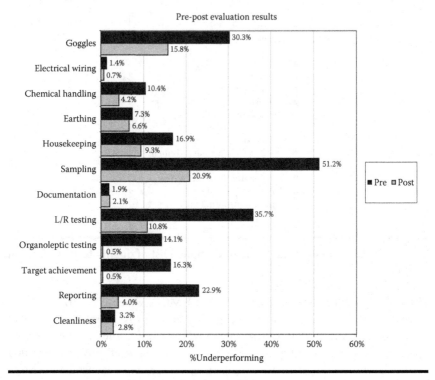

Pre-post evaluation results

%Underperforming

Figure 11.5   Nestle Pakistan Limited: Training results.

It announced that all farmers who had been supplying milk regularly for the past 15 days (i.e., March 16–31) were eligible for the scheme if they continued to supply regularly for the next 2 months. The scheme ran retroactively, i.e., it took into account the milk supplied in the 15 days of March. Farmers were given prizes in cash as well as kind through a lucky draw, with a 50% chance of winning. At the end of the 2 months, Nestle extended the scheme for another 2 months.

11.6 New Milk Processing Plants

Competition in the milk processing industry more than doubled in 2006. The first plant to start operations in 2006 was Engro Chemicals, a completely agribased company engaged in the manufacture and marketing of chemical fertilizers since 1965. Nestle felt that Engro's extensive sales network to farmers far and wide in Pakistan would provide a good springboard to them for milk collection. The other new entrant was Shakarganj Foods, part of one of the largest business groups of Pakistan, the Crescent Group. Shakarganj was previously engaged in manufacture and marketing of sugar. Shakarganj was an even bigger threat when it came to milk collection. The raw material for sugar, i.e., sugarcane, was procured in a way much similar to milk and from roughly the same geographical area as Nestle's milk shed. Shakarganj Foods enjoyed a very good relationship with sugarcane farmers and its sugarcane procurement network was heralded as the best in the sugar industry.

Nestle understood well that with increased competition, its boost in milk collection in 2005 could be short lived. For sustainable growth, the organization needed long term measures to improve the milk collection scenario. The biggest problem for milk collection was that production of milk per cattle head was low. The primary reason for this was that milk farming was not taken seriously in Pakistan. Farmers considered it a part-time activity and did not think of it in commercial terms. The whole process was managed unprofessionally and it resulted in low yield and low quality of milk. To change this situation, Nestle separated the Agri Services Group from milk collection and made the two work independently towards educating people about the benefits of modern farming techniques. The Agri Services Group was given the specific target of helping farmers and investors develop large commercial farms on modern lines where better yield would be achieved and a steady supply of milk ensured.

In addition to all that, Nestle was acutely aware of the shortage of good people in milk collection. Despite all the improvements the company made, it was still next to impossible to get good people to work in the milk collection function. Well-educated and competent people in the job market were scarce and would rather work at the head office in Lahore than in remote areas of Punjab. For instance, fresh MBAs from good universities would not agree to work in milk collection, even at Zonal Milk Collection Manager (ZMCM) level, but would be happy to join sales and marketing functions. As a matter of fact, there wasn't a single MBA in the 2200 strong milk collection organization. The quality of people in milk collection was a major bottleneck for improvement of the milk collection function.

11.7 Conclusions

Pakistan's government through its working bodies like Livestock and Dairy Development Board, Pakistan Dairy Development Company along with farmers' associations like Pakistan Agriculture and Dairy Farmers Association, Pakistan Dairy Association, and large-scale milk processors like Nestle, Engro, and Tetra Pak is striving to improve the dairy sector in the country. There are a number of projects underway. For instance, farmers' technical training program is responsible for training the farmers in farm management practices, better feeding, and milking mechanisms. Similarly, milk collection centers are being established with cooling towers. These organizations are also striving to establish farm support services like mobile veterinary hospitals, and providing the right treatment and medicines for farmers' animals. Three of these projects are given hereafter to illustrate the type of development taking place in the dairy sector in Pakistan.

11.7.1 Nestle Milk Districts

The first Nestle milk district in Pakistan was established in 1988. This was in line with Nestle's mission to improve milk collection and empower the farmers across the world in regions where Nestle had operations. Milk districts were established in Punjab, the province with highest milk production capacity in the country (see Figure 11.6). The purpose of these milk districts is to obtain fresh milk directly from the dairy producers in rural Punjab. Milk districts ensure modern milk storage facilities, dependable transportation networks, regular payments, and support services (e.g., training on good milking practices to increase milk yield for individual farmers) for farmers. In short, establishing milk districts meant signing agreements with farmers for purchase of their milk twice daily, establishing milk collection centers, installing milk chilling units, guaranteeing on-time delivery of milk to the processing plant, and ensuring that collected milk met Nestle's quality standards. In effect, the milk districts make the low cost rural production system viable where smallholder dairy producers employ mostly family labor, and rely on roughages, grasses, and crop residue for animal feed. Nestle, owing to its milk districts, purchased more than 294,000 ton of fresh milk in 2004, which increased to 379,000 ton in 2006, an increase of 29%. In 2010 Nestle Pakistan was collecting 1,040 ton of milk daily from over 140,000 farmers in about 3,500 villages of Punjab. The milk districts have helped improve the conditions of farmers in participating district. However, farmers from non-milk collecting districts have not directly benefited from this development.

11.7.2 Artificial Insemination/Crossbreeding/ Genetic Reengineering

Artificial insemination is required to increase the yield of dairy animals and to overcome the seasonal effect of buffalos' lactation cycle (see Figure 11.7). Nestle, for example, is working on developing high-yield breed of cows that would be able to survive in the hot Pakistani climate. Earlier efforts to promote imported breeds of cows had failed because those cows could not stand the climate, which is very harsh in comparison to Australia from where the partner cows had been imported. In the past few years, however, Nestle Pakistan has worked on developing a crossbreed of Pakistani, Swiss, and Australian cows that could withstand

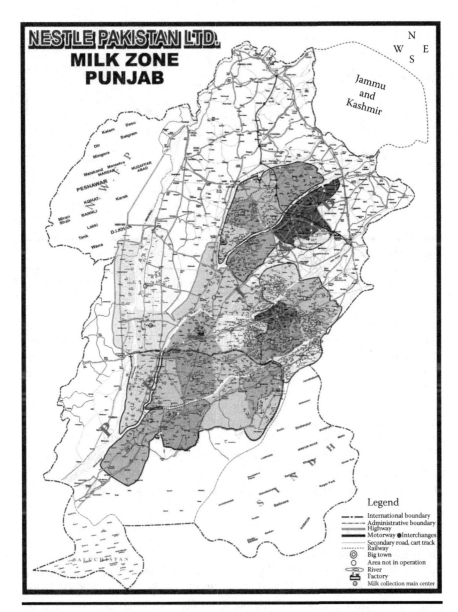

Figure 11.6   Nestle Pakistan Limited: Geographical spread of milk collection.

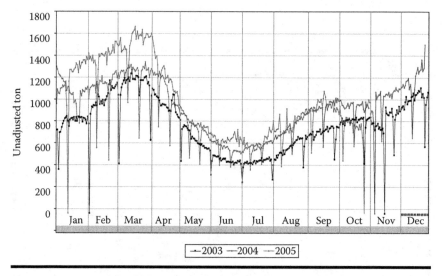

Figure 11.7   Nestle Pakistan Limited: Daily milk collection data.
Note: The dips are on every 11th of the lunar month. Many in Pakistan prefer giving
their milk out in charity, and abstain from "selling" it, on the 11th of the lunar
month.

the climate as well as deliver high volumes of milk. The first batch of these cows
arrived in Pakistan in 2008 and was crossbred with local cows. The new breed
shows better survivability in local weather; however their milk production is still
not at par with European counterparts.

Nestle is also working on use of genetics to increase milk production and
overcome the seasonal effect. The company has convinced Holland Genetics,
one of the largest dairy genetics company in the world, that milk production
in Pakistan offered an opportunity for genetics work. Consequently, Holland
Genetics has established an office in Pakistan and has started working closely
with Nestle to improve the quality and quantity of milk produced.

11.7.3 Dairy Hub

Dairy hub, a community dairy development program (CDDP), was introduced
by TetraPak, Pakistan in 2009 by building a dairy hub in Kassowal District,
Punjab. It is a concept in which farms in 20 villages located within 15–20 km
radius form a hub. Special facilities like simplified trainings, agri-services, cooling and testing equipments, and medical treatment for animals are then dedicated to a particular hub. The CDDP is aimed at developing small dairy farmers
and improving their livelihood by implementing processes to increase yield of

their animals and connect them to market channels. As a result it encourages small-scale farmers to take dairy farming as full-time commercial business.

The concept is different from Nestle's milk district model as it intends to build capacity of small dairy farmers while the purpose of milk districts is to improve collection. The CDDP intends to increase productivity at farms belonging to a hub by initiating activities like farmer trainings, breed improvement through artificial insemination, and nutrition improvement through fodder enrichment and preservation.

11.7.4 Conclusive View of What Needs to Be Done

In view of the existing scenario of the dairy sector, a lot needs to be done to bring it at par with international standards and make Pakistan one of the leading dairy exporters. First of all, plans of government and large-scale producers need to be materialized. Furthermore, the government needs to regulate the sector in order to set rules and policies regarding transactions in dairy business. Farmers' rights need to be safeguarded. In addition to policy initiatives, projects like farmer trainings in good animal husbandry and farm management practices, providing infrastructure for consistent and unadulterated supply of milk, and establishing of veterinary hospital and medical centers to provide quick and cost effective treatment of animals need to be initiated. Other projects like feed processing plants are required to provide high-nutrition feed for dairy animals at low cost. Furthermore, a farmers' cooperative that could combine resources for milk production may help in reducing per liter production cost for farmers as well as in opening up marketing channels for them and safeguarding them from exploitation of milkmen. The government, dairy sector associations, and large-scale processors should work together to encourage small farmers to improve the quantity (yield) and quality of their milk by providing the farmers with efficient dairy management tools and techniques. The government could furthermore provide credit facilities and no interest loans to small-scale farmers for enhancing the capacity of their farms.

The government and other dairy sector stakeholders like Nestle have ambitious plans for developing the dairy sector. In the end it would suffice to say that they should ensure these ideas get implemented correctly to improve the dairy sector in Pakistan.

Chapter 12

Role of Hungarian Railway on the New Silk Road

Paul Lacourbe

Contents

12.1 Introduction of Hungarian State Railways

Hungarian State Railways (Hungarian: Magyar Államvasutak or MÁV) is the Hungarian national railway company, with divisions "MÁV Start Zrt" (passenger transport) and "MÁV Cargo Zrt."

The first steam locomotive railway line started its operation in 1846, which is regarded as the origin of the Hungarian railways. During the period of the Austro-Hungarian Empire, railway operation was the responsibility of the

Hungarian government, which also inherited duties to support local railway companies. A law in 1884 provided a simplified way to create railway companies, and the development of railway speeded up. By the time the Austro-Hungarian Empire collapsed, Hungary had around 22,000 km of railway track. In particular, the company of Ábrahám Ganz invented a method of "crust-casting" to produce cheap yet sturdy iron railway wheels, which greatly contributed to the rapid development of railways in Central Europe.

The peace treaty of Trianon at the end of World War I had a devastating effect on the Hungarian railways. Since the treaty reduced Hungarian territory by 72%, the railway network was cut from around 22,000 to 8,141 km. The number of freight cars was cut from 102,000 to 27,000. As many existing railway lines crossed Hungary's new borders, most of these branch lines were abandoned. On the main lines, new border stations had to be constructed with customs facilities and locomotive service. During the peace period between world wars, development resumed and an electrification process started. However, World War II again destroyed much of the Hungarian railway system.

After World War II, the Hungarian railway network was repaired to a large extent in a short period of time. In the 1950s, accelerated industrialization gave the railway network a high priority in central planning. During the 1990s, as Hungary converted to capitalism following the fall of Berlin Wall, MÁV cut services to rural routes and focused on the profitable cargo business. In 2008, the Hungarian government sold the freight unit of the state owned company, MAV Cargo, to Austria for a total price of $593 million. Even though the opposition Fidesz party claimed that the only profitable division of the state railway had been sold off, MAV cargo has not been profitable in recent years after privatization. The sold unit actually became a client of MAV: MAV operates the rail infrastructure and leases the lines to freight operators, including the Austria-owned cargo unit.

Today, MAV remains a national company. It is a state monopoly in passenger services. The Ministry of Transport, Telecommunication and Energy exercises control on behalf of the state. Due to its national nature, the railway service has been a thorny issue during elections. In the 1990s, large-scale passenger service cuts were blocked by political pressure. In the twenty-first century, the Hungarian political establishment became very concerned with the perceived poor quality of Hungarian tracks compared to Slovakia. In 2006 the elected government promised to improve the railway network. The proposed plan by the Ministry to abandon 26 lines (or 12% of the entire network) was met with strong opposition from the local municipalities, parliamentary opposition parties, and civil organizations. On December 7, 2006, as part of a broader economic restriction package, the Hungarian government announced its intention to stop operation on 14 regional lines with a total length of 474 km, which went into effect on March 4, 2007. On April 20, 2007, the Index news web portal published material from internal MÁV studies, which indicated the new company leadership and the government intended to close all small regional railway lines after 2008 and leave only the international railway lines and large rural-to-town routes running.* However, in 2010, when Fidesz returned to power, the new government announced that they would undo a plethora of transportation decisions made by the socialists. Ten rural railway lines, previously closed with the reason of low revenues, started to operate again.

12.2 Hungary on the New Silk Road

Hungary is a landlocked country surrounded by Austria, Slovakia, Ukraine, Romania, Serbia, Montenegro, and Croatia. It is an important country geographically because it is on the crossroads between South and North, West and East, an advantage that can be further exploited in the age of globalization, if appropriate actions are taken.

One major global project impacting Hungary and the region is the so-called New Silk Road, a visionary project to reach West Europe from China by rail through the Eurasia continent. Since the New Silk Road moves from East to West, it is expected, almost surely, to reach Ukraine from Russia. It is an open question, however, how the goods will be shipped from Ukraine to Western Europe, the major market for made-in-China within the near future. Vienna is a major city in West Europe not far away from the former Iron Curtain, and is already well connected to other major cities in West Europe. Therefore, Vienna is likely to receive large amount of goods shipped from Ukraine. Currently, Hungary and Slovakia are in a fierce competition to be on the New Silk Road. For goods to arrive from Ukraine to Austria, they could travel via Slovakia or via Hungary with no major difference in time and distance. For the present moment, it appears that Slovakia is winning the competition because of better tracks and rates provided by Slovakia. In addition, other important stakeholders seem to be willing to bet on the better horse. In the most recent development, Slovakia has signed an agreement with Russia, Austria, and Ukraine for the construction of a new railway line between Ukraine and Austria via Slovakia. The line will use the same track gauge as Russia and connect the rail network of Central Europe with the trans-Siberian network, although the role of this project is not known in the broader context of the New Silk Road. This project alone will cut the freight delivery times from Europe to Asia by half, down to

---

* Wikipedia Article (http://en.wikipedia.org/wiki/Hungarian_State_Railways).

14 days from the 30 days delivery time by sea route.* Roland Berger Strategy Consultants was appointed as the main contractor to develop an investment feasibility study for the project.

Currently, Hungary is not competing against other European countries apart from the scenario mentioned. However, the route from Ukraine to Austria is only part of the big picture of the New Silk Road. Other possible routes from East to West are also being discussed. For example, China may ship the goods first to Turkey and then enter Europe from the Balkans, bypassing Ukraine altogether. China has been holding three way talks with Turkey and Bulgaria on this matter.† Should such a scenario arise; Hungary will have very different partners and competitors, especially in the Balkans.

It is noteworthy that the new EU states such as Hungary should not be viewed only as transit to the West. They are potentially very important customers as well. In fact, the New Silk Road might play a significant role in the region's transformation by boosting trade with the emerging powers in the East. From ancient times till now, Western Europe owed much of its wealth to the international sea trade. With this New Silk Road, the proximity to the economic powers in the East will favor Eastern Europe more. Interestingly, a similar development can also be expected in China. Over the years, economic development has favored Chinese regions on the sea coast that have easy access to international sea trade, while inland regions have lagged behind. With the New Silk Road, the inland regions will receive a boost in development for being on a major international trading route.

12.3 Modes of Transportation

The New Silk Road in fact refers to an Eurasia trading route. It is not limited to the railway, although the railway is likely to be the primary method of transportation, because it is faster and cheaper than highway. A highway network, passing through Kazakhstan and Orenburg region of Russia, is also envisioned. In the future, an operator will have the possibility to choose railway, highway, or a combination of the two. Actually, part of the journey may even be completed by shipping via the Black Sea, which is being discussed between China and Bulgaria.

Due to technical and political challenges as well as sheer size of project, the construction process of New Silk Road is likely to take years if not decades. One should not expect that there will be a specific day in which it is declared that this New Silk Road will be finished. Probably a finish date

---

* Russian Railways Monthly Newsletters. April 2010.
† Business Report (http://www.novinite.com/view_news.php?id = 121602).

will not exist. It is a continuous improvement and speeding up process for both highways and railways by building new infrastructures and improving the existing ones. During this lengthy process, these two modes of transportation will both compete against and complement each other. As Chinese develop the major domestic part of this international railway network, countries in East Europe are also building their own transportation infrastructure. Interestingly, China is playing an increasingly important role in such infrastructure projects. For example, China recently won a bid in Poland to construct the A2 highway connecting Warsaw with Berlin. This is the first time China entered the European Union's road construction market. As this bid occurred in the former Eastern Block, it is highly symbolic in terms of China's investment focus in the years to come.

One of the major technical problems is discrepancies in track gauge. Russia and Ukraine have a gauge size different from EU states. The newly proposed broad gauge track through Slovakia is seeking to address this issue. Apart from constructing new tracks, there are several other possibilities of trans-shipment. One method involves simply transporting goods from one wagon to another. This is likely to be the most time consuming and expensive method. If equipment and technology allow, there are two other methods available. One of them is to separate wagons from wheels and put wagons on new wheels. In a more sophisticated method, one can change the distances between wheels and get them ready for any size of gauge track. Due to the time needed for such trans-shipment, trucks are more flexible and may be justified for a relatively shorter distance.

In Hungary, there also exists competition between highway and railway, from which one can also expect similar pattern of competition on the New Silk Road. In Hungary, companies may use trucks or railway to transport their goods. Individuals may choose to take a train, a bus, or simply drive. It seems that most passengers perceive trains to be safer and faster. More importantly, the government plays a central role in this competition as the regulator. For example, the ticket price of passengers is fixed by the government. At the same time, fuel price and highway toll price are also influenced by the government. Therefore, the government decides the relative competitiveness of these two modes of transportation, with subsidies or taxes, to a large extent. In the future, EU and individual national governments are likely to influence the choices of transportation modes with similar measures.

12.4 Current Financial Situation of MAV

The current situation is not a very optimistic one for Hungary. Geographically, Hungary is not indispensable on the New Silk Road as Slovakia can largely serve as the transit country in the region. Since Hungary is a small country surrounded by many countries, which means all potential routes may bypass Hungary if the Hungarian network really turns out to be disappointing. In order for Hungary to play a significant role, it must be sufficiently competitive compared with Slovakia and other countries, in terms of equipments and contract terms (hardware and software).

MAV has been suffering losses for many years. Due to the fact that it is a state-owned company, profitability often needs to give in to political pressures. It seems very unlikely that MAV will become profitable in the near future. Even the alleged most profitable branch of MAV, the MAV Cargo, is also highly unprofitable. Actually, the buyer of MAV Cargo, ÖBB of Austria, is reportedly trying to reclaim part of the purchase price, claiming insufficient communication about the financial situation from MAV.

Currently, the major shipments passing through MAV owned infrastructure are automobiles, energy, and other heavy industries. As their production volume goes down due to the current financial crisis and deindustrialization of Europe in general, MAV freight volumes are likely to suffer as well.

Due to the financial crisis, Hungarian state as a whole is in a rather bad financial situation as large part of government revenue needs to go to reimburse EU debt. Consequently, there is almost no investment from the government into MAV. Banks are also very reluctant to provide loans as the management problem of MAV is well recognized and the banks know they cannot get returns from such investments. As investment into MAV will not bring in financial returns in the near future, the only realistic potential investors are those that will invest for strategic reasons. The two most likely such investors are EU and China.

12.5 EU Investment in Hungarian Infrastructure

Following the fall of communism in Europe, a transportation system to integrate the former eastern block is quickly envisioned. Especially after EU enlargement, the infrastructure improvement for the new EU states has been high on the agenda. Among the infrastructure frameworks, two prominent ones are the Trans-European Transport Networks (TEN-T) and the ten Pan-European transport corridors.

The TEN-T envisages coordinated improvements to the entire transportation system of Europe, in order to facilitate long-distance high-speed movement of people and freight throughout Europe. A decision to adopt TEN-T was made by the European Parliament and Council in July 1996. The projects related to TEN-T are technically and financially managed by the Trans-European

Transport Network Executive Agency (TEN-T EA), which was established exactly for this purpose by the European Commission in October 2006.*

The ten Pan-European transport corridors were defined at the second Pan-European transport Conference in Crete, March 1994, as routes in Central and Eastern Europe that required major investment over the next 10–15 years. These development corridors are distinct from the TEN-T, which is a European Union project and includes all major established routes in the European Union, although there are proposals to combine the two systems, since most of the involved countries now are members of the EU.

In both of the two frameworks, Hungary plays a very important role as a hub to connect West and East, North and South. In particular, as the Balkan states and even Ukraine aim to join EU, the bridge function of Hungary is even more important. For example, in the ten Pan-European transport corridors, three of them pass through Budapest. Among the TEN-T Priority Axes and Projects, several of them also include routes passing through Budapest. Based on existing information, EU will invest heavily into a transportation system that makes Budapest more and more important in the coming years. In August 2007, European Commission provided 7.3 billion Euros to improve the transport system of Hungary. 27.7% of the total budget is allocated to improve international accessibility to the country's rail and waterway networks. Measures include modernizing 500 km of rail tracks in Hungary. Such investment is likely to greatly improve the infrastructure of Hungary and make it more attractive for other investors such as China.

It is noteworthy that the old rival Slovakia is also high on the agenda of such EU investment, and both countries are also competing for EU funds as well, depending on which country could put forward more attractive projects. Therefore, which country would such EU investment really favor remains to be seen.

12.6 China's Investment in Hungarian Infrastructure

China follows an export-oriented growth strategy, the same strategy as Japan and Asian Tigers. It is predictable that China will continue on the same path in the years to come. In 2009, China exported $1.2 trillion of goods, was the biggest exporter in the world, and its export continues to grow rapidly. China also relies on energy and mineral import to fuel its growth. Despite the recent drive by the government to increase domestic consumption, export will remain vital for China's economic health.

At present, China's major export partners are developed countries, and China imports energy and minerals from Africa and the Middle East. That means that

---

* Wikipedia Article (http://en.wikipedia.org/wiki/Trans-European_Transport_Networks).

most of China's trade is sea bound. This makes China particularly vulnerable to eventualities in the sea, for example, military conflicts in the volatile South China Sea, terrorism, and piracy. China does not yet have a powerful navy to protect its sea lanes. This creates a rather risky situation for China. In order to overcome its reliance on the sea trade, China is ready to pay a high price, which can be seen in the recent drive to open up the New Silk Road.

As a general principle, China's overseas investment is often strategic rather than focuses on short term profit. For example, many of the infrastructure projects done in Africa do not bring in immediate financial benefits, but help to carry back the badly needed minerals. Therefore, the existing problems of MAV that deter other investors may not be as big an issue for China.

Following recent financial crisis, China, armed with trillions of dollars in reserve, is increasingly eager to look for good deals in emerging Europe where companies and governments look to reverse a decline in capital inflows by winning new foreign direct investment.* Hungary is one of such countries in need of cash. During a visit to Shanghai, the Hungarian Prime Minister Viktor Orbán said that Chinese investment in Hungary will increase significantly in the coming years. Indeed, China will play a different and increased role in the post-crisis world. China and Hungary can both benefit much from each other in a deepened partnership.† One of the issues being discussed is getting Chinese companies involved in road and railway construction projects in Hungary. The details of the projects will be worked out later by the appropriate ministries.

Actually, Hungary and China have been enjoying good relations long before the crisis. Chinese businessmen seem to prefer Budapest to other cities in the region and have established a strong presence. An office of the China Investment Promotion Agency (CIPA) has been opened in Budapest in February 2010. According to Magyar Távirati Iroda (MTI), China picked Hungary as the location for the CIPA office due to the country's central geographical location and skilled workforce, direct flights between Budapest and Beijing and the fact that Hungary is the only country in central and eastern Europe where the Bank of China is present.‡ Hungary also has a large Chinese population of around 10,000. "China sees Hungary as a regional distribution, manufacturing, and logistics center," said Hungarian Investment and Trade Development Co. Ltd. (ITD Hungary) director Gyorgy Retfalvi. Out of all potential projects,

---

* Blog by Chris Bryant on *Financial Times* (http://blogs.ft.com/beyond-brics/2010/08/23/chinese-dealmaking-in-cee/).
† Online News Report (http://www.hungarianambiance.com/2010/11/chinese-investment-in-hungary-will-grow.html).
‡ Online News Report (http://www.realdeal.hu/20100416/china-investment-promotion-agency-opens-in-budapest).

infrastructure stands out on the list. China is the factory of the world, yet the current bottleneck is the transportation route.

In summary, China really needs the New Silk Road for commercial reasons and more importantly strategic reasons, which makes China a good investor that will not seek immediate return. Hungary also stands out in the big picture because of its geographic location, which is also confirmed in the investment plan of EU. At the same time, Hungary is in urgent need for cash to reimburse its massive debt, and MAV also needs investment to become more competitive, the kind of investment that the government can not provide. China holds $3 trillion of reserves. Therefore, it is reasonable for China to invest in MAV from a strategic point of view. However, many details need to be taken care of.

One difficulty, for example, is the current structure of MAV. As it is a state-owned company, China can not deal with MAV directly, but must communicate with the ministry first, which is in charge of many other things apart from railway. This slows down the deal considerably. One solution will be to authorize MAV to deal directly with Chinese investors, with representatives from the ministry present. For China, money is not a major issue; time is much more important as Chinese are fast movers when it comes to building things. Establishing a communication expressway will be an important step.

Another difficulty relates to the nationalism of Hungarians, which may turn particularly strong during the time of crisis. The recent port deal between debt-ridden Greece and China has caused an outcry that Greece is becoming China's first colony in Europe. Similar sentiments can also be expected if a potential deal between China and Hungary is perceived to be a sale of Hungarian asset to China, and the Hungarian government may be obliged to back down due to the pressure of public opinion. One should not forget such an experience when Russians secretly bought 21.2% stake in Hungary's Oil and Gas Company MOL (Magyar Olaj-és Gázipari Nyrt).

In addition, investment in Hungary should be viewed in the broader context of China-EU relationship. One potential problem is that such investment in Hungarian infrastructure can be perceived by a few in Brussels as an attempt to break the common ground of EU states on China and exploit the complex chain of communications in EU. It will be interesting to see how EU's networks and corridors work together with China's New Silk Road.

12.7 Areas to Be Impacted by the New Silk Road

Without doubt, the New Silk Road will make it much easier for goods, services, capital, knowhow and people to move between East and West.

On the surface, it seems that China will be the biggest winner, as the "made-in-China" will increase their already very strong presence in Europe. The manufacturing sector in Hungary and the region is likely to suffer. Most importantly, factories have actually moved from West Europe to East Europe following the fall of communism. Should the massive arrival of Chinese goods force these factories to shut down, this is likely to cause agony inside EU.

However, there are also many areas in which Hungary could benefit. One of such areas would be agriculture. China is now facing a rising population, dwindling arable land and deteriorating environment. The food safety issue is high on the national agenda, a phenomenon unimaginable in Hungary. Hungary, with large size of arable land and excellent soil quality, is highly suitable for agriculture. It already has a highly developed agricultural production system that produces excellent food items, which will be easy to sell in China once the New Silk Road shrinks the delivery lead time.

In addition, the development in Hungary has not been equal across the country, with many regions left behind. With the improved infrastructure, these regions will receive a boost in development as the New Silk Road passes through them.

12.8 Conclusions

The Hungarian-Chinese relationship will be an important one for both countries. This relationship will be significantly enhanced by the forthcoming New Silk Road when people and merchandise can reach Budapest from Beijing by land in just a few days.

The current situation does not seem to be very optimistic for the Hungarian state and MAV, and investment is needed. China has much to offer financially, and China also needs to connect to the rest of the world, by land ideally. Both China and EU view Hungary as an important country logistically and are willing to bear with the less than satisfactory financial performance of MAV for a limited period of time. It is interesting to see European integration and New Silk Road go hand in hand for Hungary and the region. However, if MAV does not improve in the future, it is likely that Slovakia will fill the void. There are many things to be improved at MAV for the company to be more competitive, but it is hard to predict how much time, effort, and political will that will take. It is also not sure how the New Silk Road will impact Hungary. It seems obvious that both countries have much to offer for each other. However, whether such increased interactions will have unintended consequences remains to be seen.

SUPPLY CHAIN RISK ON THE SILK ROAD

Chapter 13

Private–Humanitarian Supply Chain Partnerships on the Silk Road

Orla Stapleton, Lea Stadtler, and
Luk N. Van Wassenhove

Contents

13.1 Introduction

In light of accelerated globalization, international supply chains are increasingly operating in volatile contexts (Simchi-Levi, 2010). Political, cultural, economic, and demographic factors render managing supply chains in multinational and diversified contexts challenging (Prahalad and Doz, 1987; Austin, 1990). This challenge is exacerbated when the company operates in regions prone to either man-made or natural disasters.

The number of natural disasters reported worldwide is increasing. Compared with 150 natural disasters reported in 1980, this rose to 540 by the year 2000. Although in recent years it has receded slightly, in 2009, over 375 natural disasters were reported (Emdat, November 2010). The number of fatalities arising from these disasters is also increasing (see Figure 13.1).

Countries along the Silk Road have traditionally been affected by disasters, be it in the form of natural disasters such as earthquakes in China, Indonesia, Thailand, and Pakistan, or in the form of conflict and civil unrest such as in India, Pakistan, the Balkans, Iran, Gaza, or Lebanon. Appendix 13.A compares the impact of natural disasters on the Americas, Asia, and Europe. Given the regional location of most Silk Road countries, we use the figures for Asia to provide a reasonable estimate of natural disasters' impact along this road. Such large-scale disasters are highly disruptive to commercial supply chains. Transportation routes, staff, warehousing depots, and other essential elements of the supply chain are often rendered non-operational after such an event. Companies operating in these countries need to be better equipped to deal with

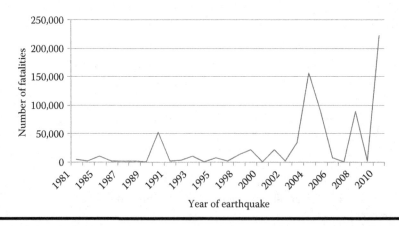

Figure 13.1 Total number of human earthquake fatalities from 301 earthquakes above 6.0 on Richter scale based on public information from the U.S. Geological Survey Earthquake Hazard program.

the challenges that come with managing multinational supply chains in a volatile region. This includes preparing to cope with unexpected disruptions caused by large-scale disasters.

International humanitarian organizations, on the other hand, are familiar with working in volatile contexts and their supply chains have been designed to respond to large-scale disasters. For example, upon receiving the European Excellence Award in 2006, a logistics award previously given only to commercial companies, the International Federation of Red Cross/Crescent Societies (IFRC) was praised as:

> They exist to operate in precisely the places where normal supply chains have broken down…despite being a global brand with relatively little direct control over its local operations, it has successfully transformed its supply chain to better meet the demands that the world places on it (Gatignon and Van Wassenhove, 2010).

However, the increasing competition for resources means that humanitarian organizations are being called on to professionalize. In this regard humanitarian organizations can learn from private sector companies. There is a high level of mutual value to be gained from these organizations working together.

This chapter is laid out in the following way. First, it provides a brief overview of the evolution of company engagement in private–humanitarian partnerships. It then compares three specific examples of large-scale disasters where companies operating supply chains along the Silk Road partnered with humanitarian organizations to assist with the disaster relief operation. They are Agility's* engagement during the Israeli-Lebanon crisis of 2006, Aramex's[†] engagement during the Gaza crisis of 2008/9, and TNT's[‡] engagement following the Asian tsunami in 2004. Each of these disaster response operations forms part of wider, ongoing partnerships between the companies and humanitarian organizations. Using these illustrative examples from the Silk Road, we examine

---

* Agility, founded in Kuwait in 1979, is a leading global provider of integrated logistics with more than 32,000 employees in 550 offices and 120 countries. Agility has built a reputation for dealing with the barriers and high-risk situations characteristic of emerging market economies.

[†] Aramex, founded in Jordan in 1982, is a leading global provider of comprehensive logistics and transportation solutions. Aramex employs more than 8100 people in over 310 locations around the world, and has a strong alliance network providing worldwide presence.

[‡] The Dutch company TNT, present in more than 200 countries with about 161,500 employees, is in charge of transferring goods and documents around the world tailored to its customers' requirements with a focus on time-definite and/or day-definite pickup and delivery.

the insights and benefits involved in partnerships between private companies and humanitarian organizations. Each company embarked on the partnership approach from a corporate social responsibility perspective but learned from it in terms of agility, operating in challenging environments, leadership development, employee motivation, and corporate identity. We illustrate how these experiences can benefit companies and help improve their commercial supply chains in non-disaster situations.

13.2 Company Engagement in Private–Humanitarian Partnerships

The last 20 years have witnessed a shift in private–humanitarian engagement. Prior to this, collaboration between the two sectors seemed unfeasible. From the humanitarian sector point of view, profit-driven companies were perceived to be the cause of, rather than solution to, problems affecting the developing world. In some cases this was tempered by the philanthropic efforts of individual companies. Overall however, the humanitarian sector dealt with companies, when necessary, on a purely commercial basis. From the private sector point of view, companies were considered to be solely accountable to their stockholders (Friedman, 1970; Margolis and Walsh, 2003). Societal issues were beyond their remit and deemed to be the responsibility of value driven organizations such as those of the humanitarian sector. Although in some cases, companies participated in philanthropic activities, generally, humanitarian organizations were seen either as potential customers or potential critics.

During the 1990s, however, a new paradigm of partnership between private enterprise and public interest emerged (Kanter, 1999). The identification of potentially lucrative markets at the bottom of the pyramid (Prahalad, 2006), combined with increasing rates of globalization, led companies to focus more on the developing world and emerging markets. Although companies' license to operate was expanding into the global space, they were increasingly being held accountable to a large set of stakeholders on whom their operations impact. The advent of the "market for virtue" (Maon et al., 2009), and the concept of companies' duty to engage for the benefit of the societies in which they operate, an approach known as "global corporate citizenship" (Schwab, 2008), meant that companies began to look for opportunities to increase their social impact (Tomasini and Van Wassenhove, 2006). Studies illustrating the link between increasing social and economic value (Porter and Kramer, 2002) also underlined the benefit to be gained from working in cooperation rather than isolation from the humanitarian sector.

Similarly, by the turn of the century, humanitarian organizations were under mounting pressure to professionalize their operations (Lindenberg, 2001).

As a growing number of stakeholders entered the humanitarian domain, the competition for funding increased. In the wake of some high profile disaster response operations towards the end of the 1990s (Samii and Van Wassenhove, 2002), donors called for greater efficiency in the use of resources. Humanitarian organizations began to recognize that the private sector could help in terms of resources and expertise (Tomasini and Van Wassenhove, 2006).

Thus, the clear cut commercial or philanthropic relationship between private sector companies and humanitarian organizations became blurred, as cross-sector partnerships were established. Figure 13.2 illustrates the resulting spectrum of engagement between private and humanitarian organizations. We focus on the gray area, that of private–humanitarian partnerships for disaster response operations.

Company contributions which have the highest impact on the social sector use the core competencies of the business (Kanter, 1999). In terms of disaster relief operations, the supply chain and logistics functions are crucial for the operation's success (Van Wassenhove, 2006). Companies have the supply chain management capability to enhance the efficiency of the relief supply chain. On the other hand, companies can learn from the humanitarian sector, in terms of agility, adaptability, and innovation in difficult circumstances (Samii et al., 2002; Van Wassenhove et al., 2008). Partnerships between companies and humanitarian organizations have the potential to deliver three key benefits. They are: fast, effective support during a disaster; capacity building between disasters; and, exchange of ideas and best practices between sectors (Van Wassenhove et al., 2008). However, developing successful partnerships can be challenging due to a variety of reasons. In the first place, a degree of suspicion still exists between the two sectors. Second, companies generally operate at a different clock speed to humanitarian organizations. Furthermore, the areas of expertise of each sector need to be well coordinated to avoid duplication.

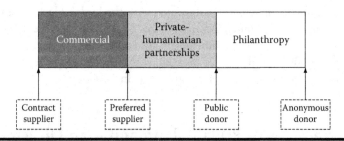

Figure 13.2 The private–humanitarian engagement spectrum. (From Tomasini, R. and Van Wassenhove, L.N., Overcoming the barriers to a successful cross-sector partnership, *The Conference Board Executive Action Series*, p. 188, 2006.)

The three examples of private-partnerships in this chapter involve global logistics companies. In each case, their approach was largely driven by the companies' top management. As Agility's chairman and managing director, Tarek Sultan, explained to every employee:

> The issues of poverty, disaster, and disease are ones that no one can afford to ignore. Given our expeditionary logistics capabilities and structural network on the ground throughout the world, we are in a unique position to help (Tomasini and Van Wassenhove, 2009).

Aramex's community engagement was also driven by the philosophy and motivation of CEO and founder Fadi Ghandour. As he stated:

> A company's development is intertwined with the well-being of all stakeholders. Therefore, it is imperative for us to take on an activist approach (Aramex press release, January 2010).

In the case of TNT, the leading question of a Business Week article "What are you doing for the world after September 11th?" had a fundamental impact on CEO Peter Bakker, who concluded that TNT had a responsibility to act as a good global corporate citizen (Tomasini and Van Wassenhove, 2004). The companies' customers and employees also demanded increased social engagement, which reinforced the top management's conception of corporate citizenship. Each of the companies has embarked on long-term partnerships with organizations from the humanitarian sector.

In the following section, we illustrate examples of three disasters along the Silk Road, where these companies' partnerships with humanitarian organizations were used to great benefit in the relief operation. The first example is Agility's partnership with the Kuwaiti Red Crescent Society to deliver relief supplies to those displaced by the Lebanon crisis in 2006. Second, Aramex initiated a broad donation campaign and cooperated with the Jordan Hashemite Charity Organization and the United Arab Emirates (UAE) Red Crescent Society to deliver donations to the communities in need following the Israeli attacks in Gaza in 2008/09. Third, TNT engaged with the World Food Programme (WFP) in the aftermath of the tsunami in Asia in 2004.

13.3 Examples of Private–Humanitarian Disaster Relief Partnerships on the Silk Road

Having a large customer base along the Silk Road as well as substantial logistics capabilities on the ground, each of the three companies felt a duty to engage in disaster relief operations in the region. For example, when the Israeli army fired missiles into Lebanon in retaliation for attacks by Hezbollah, Agility's employees called for the company to get involved in the relief operation (Tomasini and Van Wassenhove, 2009). They argued that Agility's valuable logistics expertise and structural network could provide essential help for humanitarian initiatives. The fact that at least seven of Agility's 120 employees in Lebanon had been displaced and several clients had lost warehouses or had operations severely disrupted, strongly supported the argument in favor of company engagement (Tomasini and Van Wassenhove, 2009).

Aramex on the other hand was asked by a very close business partner, a communications company, to use their logistics capabilities for the benefit of people in need in Gaza (Stadtler and Van Wassenhove, 2010). Finally, in light of the long-term partnership with WFP and the company's local presence in the area, TNT felt committed to engaging in the relief operation in response to the Asian tsunami. Considering each company's logistics expertise, contributing to the relief supply chain of a humanitarian organization was deemed the best way to enhance the relief operations. In each case, the companies relied on their humanitarian partners' valuable disaster relief expertise and capacity during the engagement.

13.3.1 Disaster Relief Operation

Companies can contribute in various ways and at different stages of the humanitarian relief supply chain. This includes: specialized resources and expertise; sharing physical logistics resources (e.g., airplanes, trucks, warehouses); financial donations or donations of company products (e.g., food and relief equipment); secondment or allocation of staff; and, access to organizational capabilities and resources (e.g., tracking and routing systems) (Thomas and Fritz, 2006). To date, Aramex, Agility, and TNT have been involved in multiple disaster relief operations. Depending on the disaster's scope and humanitarian organizations' needs, they have contributed to almost all stages of the relief supply chain. We focus on the three specific disasters along the Silk Road to illustrate examples of implementing private–humanitarian partnerships for disaster relief operations (see Figure 13.3).

The first step in the relief supply chain, *planning*, takes place before a disaster occurs and consists of developing contingency plans and procedures to respond to disasters (Van Wassenhove et al., 2010). At this stage, humanitarian organizations coordinate their plans with governments of potentially affected countries. While companies rarely intervene in this planning stage, they can play a major role in improving a humanitarian organization's preparedness for a disaster. In fact, as Jorge Olague, donor relations officer at WFP, explained,

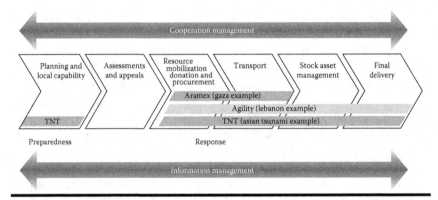

Figure 13.3 Company engagement in the humanitarian relief supply chain in three exemplary disasters. (Based on Van Wassenhove, L.N. et al., An analysis of the relief supply chain in the first week after the Haiti earthquake, 2010, www. insead.edu/facultyresearch/centres/isic/documents/HaitiReliefSupplyChain_ Final25Jan.pdf)

> Any dollar given before an emergency goes much further than more dollars given after (quoted in Maon et al., 2009).

In 2002, about 3 years before the tsunami disaster, TNT and WFP established the "Moving the World" partnership. Under the partnership umbrella, joint projects concentrated on increasing preparedness for more effective disaster responses. They addressed WFP's common logistics needs and efforts and sought to improve WFP's warehouse management, fleet management, and procurement strategies. Furthermore, they carried out a project to reshape WFP's accounting, auditing, and human resource management. Finally, TNT helped WFP diversify their donor base with new fundraising strategies and business plans (Samii and Van Wassenhove, 2004). These actions facilitated WFP's preparedness to respond to disasters such as the Asian tsunami in 2004.

In the aftermath of a disaster, the immediate response of humanitarian organizations is to carry out an *assessment* within the first 72 h. This assessment considers the magnitude of the damage, the availability of resources for the response, as well as the immediate needs of the victims. Simultaneously, humanitarian organizations prepare formal *appeals* to support their relief operation. Companies rarely engage in this stage as they generally lack the necessary expertise and networks. Furthermore, the time pressure limits the humanitarian organizations' capacity to collaborate during this assessment and appeal stage.

In the subsequent *resource mobilization, donations, and procurement* stage, company engagement can be of significant value. It is precisely at the start of a disaster relief operation that humanitarian organizations often lack access to resources in terms of logistic capacity. Consequently, in the example of the Asian tsunami, TNT allocated staff, trucks, planes, and material on the ground. Furthermore, companies can help mobilize and collect donations—be it financial or in-kind donations. Financial donations are used by humanitarian organizations to procure warehouse space and housing in the field, in addition to equipment and relief items. For example, TNT mobilized financial donations from their employees in the Asian region and from a fundraising campaign called "Just one hour can make a difference" at TNT Netherlands. During this campaign, TNT encouraged employees to work one extra hour and then donated the hourly wage to WFP (TNT, 2005). Furthermore, TNT assisted the donation process by setting up drop-off points for donated items.

Aramex initiated a broad media campaign to call for donations (Stadtler and Van Wassenhove, 2010). They selected appropriate collection points and applied their competencies and capacity to collect the donations, prepare them for transportation to their warehouses, and hand them over to the Jordan Hashemite Charity Organization and the UAE Red Crescent Society. Aramex received offers for help from many corporate and charitable partners who helped spread the word and contributed via donations, volunteer workforce, trucks, or other operational assets. Finally, during the Lebanon relief operations, Agility leveraged resources in terms of transport and warehouse capacities, human resources, as well as contacts and networks (Tomasini and Van Wassenhove, 2009).

The *transport* stage aims to deliver the aid to the right place at the right time. In this stage, companies can contribute their staff expertise, local knowledge, and transport equipment. For example, TNT helped WFP transport material to the affected areas. They formed an air operations ramp management team that supported the loading and unloading of disaster response flights. Furthermore, they organized a convoy of trucks and chartered two helicopters for transportation (TNT, 2005). With regard to Aramex and Agility, they both helped their respective humanitarian partners with trucks and staff in the transport stage.

Relief material is stored in warehouses that need to be located strategically to use the available infrastructure and, at the same time, guarantee safety conditions for assets and people. Companies can support this *stock asset management* process with their networks and logistics capacity. For example, during the Lebanon crisis, there was a strong need for warehousing solutions in Syria, close to the Lebanese border. Using their contacts, Agility identified warehouse owners and transport operators in Syria. Then, they visited the warehouses and inspected the trucks. Within a short time, they managed to set up a solid regional transport and storage network for the humanitarian organizations (Tomasini and Van Wassenhove, 2009). TNT also engaged in the stock asset management stage in the aftermath of the tsunami. Not only did they help WFP plan the required transportation and storage, but they also offered them office space for a control team and storage space for relief items (TNT, 2005).

The last stage in the relief supply chain, *final delivery,* involves last mile distribution of relief items to the beneficiaries. Constraints within the humanitarian organization's resources call for companies' support during this stage also. For example, the Kuwaiti Red Crescent Society faced significant challenges in Lebanon since they lacked any local bases, resources, or logistics on the spot.

> "The Lebanese borders were closed and all crossing points had been bombed. We were having difficulties getting drivers into Lebanon and finding warehouses. The only way to find solutions was to work with the locals who knew the roads (and) could assess tensions (...)," as the Kuwaiti Red Crescent Emergency Relief Coordinator, Mohamed Hassan, explained (Tomasini and Van Wassenhove, 2009).

Thus, Agility put together a team in Lebanon which ensured that the supplies traveling the last mile would reach the intended beneficiaries. Likewise, TNT in Asia made staff available to help coordinate warehousing and final delivery (TNT, 2005).

Figure 13.3 provides an overview of the companies' engagement along the humanitarian relief supply chain in these specific examples. Aramex contributed significantly during the resource mobilization, donations, and transport stages. Agility mainly engaged in the resource mobilization, transport, stock asset management, and final delivery stages since they had valuable contacts, expertise, and resources on the spot. Finally, in the aftermath of the tsunami, TNT helped WFP during the planning, resource mobilization, donations, transport, stock asset management, and final delivery stages. These examples demonstrate that there is no single best way for companies to engage in the humanitarian supply chain. Rather, their engagement is driven by matching the humanitarian organizations' requirements with the company's global and local resources, networks, and established partnerships on the ground.

13.3.2 Partner Interdependence

The previous section illustrated how companies can contribute to the humanitarian relief supply chain. We now draw our attention to the points of interaction between the partners in each of the described private–humanitarian partnerships. A successful partnership requires clarification on how the partners will work together in terms of distribution of tasks and responsibilities and how they will coordinate their activities. The extent to which partners work together and carry out several or one specific supply chain step together is influenced by their interdependence.

In sequential interdependence one partner hands over its resources, products, or finalized services to the other partner. The partners coordinate to ensure that the services or products comply with the requirements and to allow a smooth transfer at the interaction point. Aramex collected and forwarded donations to the Jordan Hashemite Charity Organization and the UAE Red Crescent Society. These organizations had access to Gaza and also the latest information about what items were allowed to cross the border (Stadtler and Van Wassenhove, 2010). Thus, they were able to keep up with the continuously changing requirements of the local authorities, a significant challenge for the operation. Aramex invited one representative of the Jordan Hashemite Charity Organization to work in the company's coordination team. This key person as well as the coordination team as a whole was crucial to provide the packing teams with the necessary information. Furthermore, Aramex discussed in detail with the two humanitarian organizations, how to hand over the items and ensure an efficient subsequent course of the supply activities.

In reciprocal interdependence, partners exchange and coordinate resources and outputs. Given the various points of interaction, partnerships here generally need greater and more complex coordination mechanisms (Gulati and Singh, 1998; Gulati et al., 2005). Agility's engagement in Lebanon shows a reciprocal interdependence with the Kuwaiti Red Crescent as they set up a field team at the Syrian-Lebanese border to serve as liaison. Only this way could the Kuwaiti Red Crescent use a safe route across the Lebanese border. Similarly, the TNT staff in Asia worked hand in hand with WFP staff on the ground. To achieve the most effective outcome, the partners were dependent on each other. Such a reciprocal interdependence requires effective communication, joint processes, and mutual adaptation. The companies' engagement and the partners' interdependence in the three relief operation examples are summarized in Table 13.1.

13.4 Lessons from Private–Humanitarian Partnerships

Each private–humanitarian partnership described here significantly contributed to improving the disaster relief operation. The companies drew concrete lessons from these examples, and through repeated involvement in disaster relief operations have consistently improved their approach to private–humanitarian partnerships. In this section, we discuss the major lessons the companies have learned with regard to the preparation and implementation of these partnerships.

Table 13.1   Comparison of Companies' Engagement in Private–Humanitarian Partnerships on the Silk Road

The Disaster Operation	Agility (Lebanon Example)	Aramex (Gaza Example)	TNT (Asian Tsunami Example)
Company engagement in the humanitarian relief supply chain	Resource mobilization Donations Transport Stock asset management Final delivery	Resource mobilization Donations Transport	Planning Resource mobilization Donations Transport Stock asset management Final delivery
Private–humanitarian partner interdependence	Reciprocal	Sequential	Reciprocal

13.4.1 Partnership Preparation

The success of private–humanitarian partnerships largely depends on the quality of planning and preparedness involved. Preparation within a company starts with an evaluation of potential value of the company's contribution, the resulting company benefits, and the risks implied. Engagement in disaster relief operations usually involves high risks due to the contextual uncertainty. Along the Silk Road, disaster relief operations frequently have to cope with limited local capacities and an already underdeveloped local infrastructure. The UN Human Development Index ranks Silk Road countries such as China, India, Pakistan, and Afghanistan as having medium to low human development, while countries such as Iraq and Lebanon have suffered from protracted conflicts that have had a significant impact on the country's infrastructure (UNDP, 2010).

Once a company has decided to engage in humanitarian partnerships, they need to find a complementary partner. Having conducted the necessary due diligence, TNT chose WFP. They filtered candidates with the right organizational focus according to reputation, neutrality, and global presence. They narrowed the focus based on four weighted criteria: matching competencies, public relations values, effectiveness and overhead costs, and geographical scope. Finally, TNT compared the remaining candidates on their organizational fit (Tomasini and Van Wassenhove, 2004).

In response to their ad-hoc engagement in Lebanon, Agility saw the necessity to establish long-term partnerships with selected humanitarian organizations. They formulated precise selection criteria in terms of legal registration, neutrality, and impartiality in service delivery. Additionally, a prior relationship as well as a common logistics background or common regional or local knowledge were favored characteristics (Tomasini and Van Wassenhove, 2009). Agility now builds on their long-term partnership with the International Medical Corps and participates in the Logistics Emergency Teams* (LETs) to foster disaster preparedness. These long-term and preparedness enhancing partnerships play an important role in building capacity and improving disaster preparedness for both partners by nurturing sound relationships and defining the respective roles and responsibilities.

Crucial for the success of partnerships with humanitarian organizations is sound support within a company. This helps to overcome potential employee resistance. In this respect, TNT organized joint events with WFP, encouraged company-internal communication on the topic, and broadly explained the rationale behind TNT's humanitarian engagement to employees (Samii and Van Wassenhove, 2004). Since disaster responses such as the tsunami operation in 2004 are organized in a decentralized way, internal support in terms of employee motivation and empowerment are crucial.

To prepare for a specific disaster relief operation, companies benefit from having clear criteria, guidelines, and procedures for intervention. For example, Agility made the decision that they would only engage in countries where they have an office or local capacity (Tomasini and Van Wassenhove, 2009). Existing relationships with transport operators and customs officials, access to local warehousing and transport assets, and local expertise are considered essential to offer support. The Lebanon engagement also encouraged Agility to develop a sophisticated Humanitarian and Emergency Logistics Programme (HELP) within the company to guide further engagement with humanitarian organizations (Tomasini et al., 2009).

Agility decided not to get involved if the security situation was worse than a UN code 3. Likewise, they developed a two-pronged approach for natural disasters that distinguishes between major and small ones (Tomasini and Van Wassenhove, 2009). Furthermore, clear terms for disaster relief

---

* The Logistics Emergency Teams pool logistics expertise, human resources, and in-kind services of the three global logistics companies TNT, Agility, and UPS. These resources are made available to the humanitarian community when a disaster strikes. The support of LETs includes providing logistics specialists (e.g., airport coordination, airport managers, and warehouse managers), logistics assets (e.g., warehouses, trucks, forklifts), and transportation services.

operations have been defined. They include: standard operating procedures, a disaster relief kit, and decision-making capacity for Agility country offices. This facilitates immediate action in response to a disaster (Tomasini and Van Wassenhove, 2009). Since the Lebanon engagement, Agility has contributed to response operations of at least 20 disasters and has continuously improved their approach.

To successfully build on a partnership for disaster relief operations, it is important for staff to be involved in and equipped for the planning, decision-making, and implementation processes. Risks occur if staff members lack adequate expertise, capacity, and necessary resources. Likewise, without their inclusion in decision-making, compliance tends to be weak (Harley and Warburton, 2008; Van Wassenhove et al., 2008). In this respect, TNT offers their employees intensive briefing and de-briefing sessions as well as workshops throughout its long-term partnership with WFP (Tomasini and Van Wassenhove, 2009). In line with their HELP program, Agility now requires that volunteers are trained in first aid, humanitarian issues, and ethics (Tomasini and Van Wassenhove, 2009). Similarly, Aramex has developed clear guidelines and codes of conduct for volunteering employees. The challenge is to prepare guidelines and frameworks that give employees orientation and set direction and, at the same time, to provide employees with enough discretionary power to shape operations and develop ownership (Stadtler and Van Wassenhove, 2010).

13.4.2 Implementation Challenges and Solutions

Partnering for disaster relief operations is challenging. Partners sometimes work together for the first time, demand for relief tends to be unpredictable, local infrastructure is often destabilized, transport capacity is limited, political complexities are intense, and information is fragmented and hard to interpret (Kelly, 1995). In this context, engaging in partnerships for disaster relief along the Silk Road involves a substantial risk of harm to staff, property, and/or reputations. Equipment and other property are often not covered by insurance. There is thus a great need to ensure that there is a planning and coordination team in charge that reduces risks as much as possible.

While the risk of damage to equipment is clearly costly, the hidden cost of harm done to reputation if the partnership fails, or in case of misconduct, is also critical (Harley and Warburton, 2008). For example, during Aramex's donation campaign for Gaza, there was the risk of harm to their reputation if the campaign's message was misunderstood. However, the fact that relief operations are subject to the principles of humanity, neutrality, and impartiality (Tomasini and Van Wassenhove, 2009) reinforce the non-political nature of a company's involvement. Equally, during the operation's implementation, there is a risk of mistakes that can result in injuries to employees and volunteers (Stadtler and Van Wassenhove, 2010). Partners need to make sure that employees are well informed about the latest state of the disaster situation and not overworked.

Each company also took steps to ensure their humanitarian engagement was well coordinated with their ongoing core business operations. Agility decided to exclude employees working on bidding and contract relations with humanitarian organizations from company disaster response activities (Tomasini and Van Wassenhove, 2009). Aramex showed that potential tensions stemming from the coordination with the core business can be approached by separating the business and disaster relief activities either geographically (e.g., using different warehouses) and/or on a time basis (incorporating different work shifts) (Stadtler and Van Wassenhove, 2010).

During their humanitarian efforts, Agility, Aramex, and TNT realized the necessity of continuous direct communication in light of the time pressure and changing contextual factors. Within the partnerships, clear planning and coordination was important to leverage each organization's strengths. At the same time, partners needed to be flexible and open to constant adaptation. As one way to ensure flexibility as well as effectiveness and safety, the companies provided local staff with discretionary power as long as they fulfilled the requirements.

Finally, private–humanitarian partnerships combine different organizational backgrounds, ways of working, and corporate cultures. Challenges may arise within a partnership, such as lack of mutual understanding, lack of transparency and accountability, unbalanced commitment, and weak relationship management (Tomasini and Van Wassenhove, 2006). The practical examples show that these differences come to light particularly in long-term partnerships. For example, TNT and WFP realized that, although they both work on logistics, they each have their own approach to it—just as they have different organizational goals and objectives (Tomasini and Van Wassenhove, 2006). While the company's objectives are related to costs, humanitarian objectives tend to be more centered on satisfying beneficiary needs (Tomasini and Van Wassenhove, 2006). Furthermore, decision-making at TNT was less bureaucratic and politically sensitive than at WFP (Samii and Van Wassenhove, 2004). Both TNT and WFP learned that communication and specifying each partner's needs are crucial to develop a mutual understanding. The best time to do so was during non-emergency times when pressure was less intense (Tomasini and Van Wassenhove, 2006). Table 13.2 summarizes the lessons from corporate engagement in private humanitarian partnerships.

Table 13.2 Insights from Company Engagement in Private–Humanitarian Partnerships

Main Drivers:	CEO Leadership	Employee Demands	Customer Demands	
Partnership: preparation	Sophisticated partner selection process and prior relationships between partners	Sound internal support	Development of clear criteria for engagement, guidelines, and standard operating procedures	Staff involvement and training
Implementation: Challenges	Risk of harm to staff, property, and/or reputation	Potential conflict of interest with core business	Allowing flexibility while ensuring effectiveness and safety	Partner differences in terms of working cultures and objectives
Implementation: Solutions	Importance of sound planning and coordination processes	Geographical, time-based, or employee-based separation	Local decision-making and continuous communication	Open communication and development of mutual understanding between disasters

13.5 Benefits of Private–Humanitarian Partnerships to Companies

As we stated at the beginning of this chapter, accelerated globalization has two key implications for companies. First, multinational companies have a license to operate in the global space. This increasingly includes volatile regions, such as along the Silk Road. According to the World Disaster Report 2010, the six largest disasters in 2009 affected approximately 100 million people. Of these six disasters, five occurred in Asia and affected 80 million people. Twelve other major natural disasters occurred that year affecting between one and six million people. Of these 12 disasters, 9 occurred in Asia (IFRC, 2010). Second, companies' accountability in the regions they operate is expanding to include multiple stakeholders. In an effort to increase economic value while increasing social value, many companies are looking to engage with the public or humanitarian sectors.

In order for the partnership to be of real benefit to both humanitarian organizations *and* companies, it is important that the core competencies of each partner are well aligned, and that the partnership is well managed. The supply chain management expertise of global logistics firms can help to increase the capacity of humanitarian organizations at different and multiple stages of the humanitarian relief supply chain. The three examples of partnerships for disaster relief operations along the Silk Road in this chapter show how companies can contribute effectively. At the same time, managing private–humanitarian partnerships is challenging due to a number of reasons including: lack of mutual understanding, different organization/cultural backgrounds, or risk to personnel and equipment. However, as the three examples show, recognition of the organizations' interdependence combined with communication, coordination, planning, and preparedness prior to the disaster can help mitigate the effects of these challenges.

As a result of their humanitarian engagement, each company drew some important benefits for their core business. The examples show that engagement in partnerships for disaster relief teaches companies how best to deal with supply chain disruptions, damaged or completely destroyed infrastructure, and high levels of uncertainty. As one way to ensure flexibility and speed of operations, the companies empowered those on the ground with discretionary power, the necessary knowledge, and the capacity to make operational decisions.

Furthermore, the volatile context of the relief operations engagement gave the employees the opportunity to enhance their leadership skills. Decisions had to be made quickly, responsibility had to be taken immediately, and employees had to work closely together. Indeed, all three examples showed that a well functioning team is crucial during a disaster relief operation. These skills—improved leadership, team work capabilities, and adaptive approaches—are important to companies dealing with the challenges of managing multinational supply chains in volatile regions. As the Silk Road is frequently impacted by disasters, lessons from humanitarian engagement will help companies operating in the region to develop their response strategies and safeguard against the disruptive impact of disasters (Binder and Witte, 2007).

Private–humanitarian partnerships are a way to gain insight into managing challenging environments that are typical of many countries along the Silk Road. They also deepen the company's ties with key stakeholders in the region. In the case of Aramex, their collaborative engagement let them deepen ties, not only with the humanitarian partners but also with local authorities and various business partners who supported the response campaign (Stadtler and Van Wassenhove, 2010). This can enable companies operating in volatile regions to improve their capacity even in non-disaster settings. As CEO of Agility, Tarek Sultan reflected:

> We work in different countries in challenging situations; therefore we need to be ready to add value in tough places, whether it is on commercial terms or not. We understand risk, and we have an appetite and knowledge to thrive on it and do business. (...) When you are sitting in Kuwait, you understand that everything can change overnight and affect not only your business but people around you (Tomasini and Van Wassenhove, 2009).

Besides the operational learning, humanitarian engagement also has a positive impact on the company's image. In fact, humanitarian engagement can help build up brand equity, establish relations with external stakeholders (Binder and Witte, 2007), and gain access to new markets. With regard to customers, engagement in these types of operations boosts the company's reputation as a reliable service provider in stressful situations. This is particularly relevant for a logistics provider operating along the Silk Road.

Internally, private–humanitarian partnerships can also facilitate a company's access to new talent or motivate existing staff. Employee surveys among TNT have shown that at least half of their employees have been involved in the partnership with WFP and about two thirds of the employees find that the partnership makes the company more attractive to work for (Van der Kaaij and Hooijberg, 2006). Increasing employee satisfaction and loyalty brings multiple benefits to the company, such as: reducing the cost of staff turnover and retaining institutional knowledge.

To conclude, when designed and managed correctly, private–humanitarian partnerships have mutual benefits for those involved. Humanitarian organizations benefit from resources and professional experience. Companies also benefit from humanitarian expertise and improve in terms of agility, operating in challenging environments, leading teams, improving familiarity with the region, and deepening relationships with key stakeholders. Figure 13.4 illustrates the key benefits for each partner. With the rise in natural disasters and ongoing political instability worldwide, these learning experiences will be increasingly valuable to companies operating in volatile regions, such as along the Silk Road.

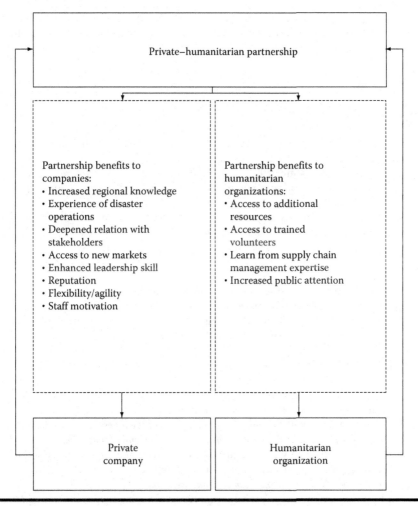

Figure 13.4 Benefits of private–humanitarian partnerships to companies and humanitarian partners.

13.A Appendix: Natural Disasters

Total 2000–2009	Americas	Asia	Europe
No. of natural disasters	1,334	2,903	996
No. of fatalities	32,577	933,250	91,054
No. of people affected (in thousands)	73,161	2,159,715	10,144
Estimated damage (millions of US$)	428,616	386,102	146,414

Source: From EM-DAT, CRED, University of Louvain, Belgium in: International Federation of Red Cross and Red Crescent Societies (2010): World Disaster Report 2010.

References

Austin, J. E. 1990. *Managing in Developing Countries: Strategic Analysis and Operating Techniques.* The Free Press, MacMillan Inc., New York.

Binder, A. and J. M. Witte. 2007. *Business Engagement in Humanitarian Relief: Key Trends and Policy Implications.* Humanitarian Policy Group, Overseas Development Institute, London, U.K.

Friedman, M. 1970. The social responsibility of business is to increase its profits. *New York Times Magazine,* September 13, pp. 32–33, 122, 124, 126.

Gatignon, A. and L. N. Van Wassenhove. 2010. The Yogyakarta earthquake: Humanitarian relief through IFRC's decentralized supply chain. *International Journal of Production Economics,* 26(1):102–110.

Gulati, R., P. Lawrence, and P. Puranam. 2005. Adaptation in vertical relationships: Beyond incentive conflict. *Strategic Management Journal,* 26(5):415–440.

Gulati, R. and S. H. Singh. 1998. The architecture of cooperation: Managing coordination costs and appropriation concerns in strategic alliances. *Administrative Science Quarterly,* 43(4), 781–814.

Harley, M. and J. Warburton. 2008. Risks to business in social involvement. *Journal of Corporate Citizenship,* 29:49–60.

Kanter, R. M. 1999. From spare change to real change: The social sector as beta site for business innovation. *Harvard Business Review,* May–June, 77(3):122–132.

Kelly, C. 1995. A framework for improving operational effectiveness and cost efficiency in emergency planning and response. *Disaster Prevention and Management,* 4(3):25–31.

Lindenberg, M. 2001. Are we at the cutting edge or the blunt edge? Improving NGO organizational performance with private and public sector strategic management frameworks. *Non-Profit Management and Leadership,* 11(3):247–270.

Maon, F., A. Lindgreen, and J. Vanhamme. 2009. Developing supply chains in disaster relief operations through cross-sector socially oriented collaborations: A theoretical model. *Supply Chain Management: An International Journal,* 14(2):149–164.

Margolis, J. D. and J. P. Walsh. 2003. Misery loves companies: Rethinking social initiatives by business. *Administrative Science Quarterly*, 48:268–305.

Porter, M. E. and M. R. Kramer. 2002. The competitive advantage of corporate philanthropy. *Harvard Business Review*, December, 80(12):57–68.

Prahalad, C. K. 2006. *The Fortune at the Bottom of the Pyramid*. Wharton School Publishing, Upper Saddle River, NJ.

Prahalad, C. K. and Y. L. Doz. 1987. *The Multinational Mission*. Collier Macmillan Publishers, London, U.K.

Samii, R. and L. N. Van Wassenhove. 2002. *IFRC Choreographer of Disaster Management: Preparing for Tomorrow's Disasters—Hurricane Mitch*. INSEAD case study no. 02/2008-5039.

Samii, R. and L. N. Van Wassenhove. 2004. *Moving the World: The TPG Partnership-Learning How to Dance*. INSEAD case study no. 02/2008-5194.

Samii, R., L. N. Van Wassenhove, and S. Bhattacharya. 2002. An innovative public-private partnership: New approach to development. *World Development*, 30:991–1008.

Schwab, K. 2008. Global corporate citizenship. *Foreign Affairs*, 87(1):107–118.

Stadtler, L. and L. N. Van Wassenhove. 2010. *Corporate Social Engagement: How Aramex Crosses Boundaries*. INSEAD case study, ECCH reference no 711–038-1.

Thomas, A. and L. Fritz. 2006. Disaster relief incorporated. *Harvard Business Review*, 84(11):114–122.

Tomasini, R. and L. N. Van Wassenhove. 2004. *Moving the World: The TPG Partnership-Looking for a Partner*. INSEAD case study no. 02/2008-5187.

Tomasini, R. and L. N. Van Wassenhove. 2006. Overcoming the barriers to a successful cross-sector partnership. *The Conference Board Executive Action Series*, p. 188.

Tomasini, R., M. Hanson, and L. N. Van Wassenhove. 2009. *Agility: A Global Logistics Company and Local Humanitarian Partner*. INSEAD case study no. 10/2009-5559.

Van der Kaaij, J. and R. Hooijberg. 2006. *Corporate Philanthropy at Work: U2 Can Move the World (B): Mailmen on a Mission*. IMD case study no. 2-0126.

Van Wassenhove, L. N. 2006. Blackett memorial lecture humanitarian aid logistics: Supply chain management in high gear. *Journal of the Operational Research Society*, 57(5):475–489.

Van Wassenhove, L. N., R. Tomasini., and O. Stapleton. 2008. Corporate responses to humanitarian disasters: The mutual benefits of private-humanitarian cooperation. Conference Board Research Report R-1415-08-WG.

Websites

Aramex. 2010. Learning grassroots logistics, an Aramex white paper on the "Deliver Hope to Gaza" relief campaign. www.aramex.com/content/uploads/100/55/35297/Gaza%20white%20paper-%20Jan%2017.pdf

TNT. 2005. TNT support for disaster relief efforts in Asia, press release, 04.01.2005, http://group.tnt.com/press/400477/TNT_support_for_disaster_relief_efforts_in_Asia.aspx

Van Wassenhove, L. N., A. Pedraza Martinez., and O. Stapleton. 2010. An analysis of the relief supply chain in the first week after the Haiti earthquake. www.insead.edu/facultyresearch/centres/isic/documents/HaitiReliefSupplyChain_Final25Jan.pdf

International Federation of Red Cross and Red Crescent Societies. *World Disaster Report, 2010.* http://www.ifrc.org/Global/Publications/disasters/WDR/WDR2010-full.pdf

United Nations Development Programme. *Human Development Report, 2010.* http://hdr.undp.org/en/media/HDR_2010_EN_Table1_reprint.pdf

www.agilitylogistics.com

www.aramex.com

www.tnt.com

www.wfp.org

Conference Presentation

Simchi-Levi, D. 2010. Creating value in a volatile world. ALIO–INFORMS Joint International Meeting Plenary Presentation, Argentina.

Chapter 14

Incorporating Harvest, Maturity, Yield, and Demand Risk in Planning for Agricultural Supply Chains for Premium Products

Barış Tan

Contents

14.1 Introduction

Utilizing agricultural resources in the best way possible by increasing agricultural productivity has become one of the top priorities in the agricultural sector. Managing agricultural resources effectively allows countries to improve their economies and increase their income, provide food and nutrition to their population by relying on domestic sources. In the last century, improvements in global agricultural productivity were not realized equally in all countries. With large populations, vast agricultural resources that are faced with natural risks, countries placed along the Silk Road can benefit significantly from effective management of agricultural supply chains.

Advances in science and technology allowed improvements in soil, water, nutrient, and pest management while crop production improved significantly with the advances in biotechnology. These improvements increased yield and reduced environmental impact of agricultural supply. However, agricultural supply is still prone to risks associated with the biological nature of production process and risks associated with weather. Changes in weather conditions affect the maturation time for seeds and the start of the harvest time. An unexpected weather condition such as a cold front or a heavy rain may end the harvest season at an unexpected time. Attack of harmful moths may decrease the yield from a farm. Combined with long lead times and not being able to increase the capacity to meet unexpected demand, supply uncertainty is quite prevalent in agricultural supply chains.

In recent years, premium agricultural products have begun to secure a more important part in grocery retail markets. This development was led by increasing number of customers who prefer and pay a premium for differentiated agricultural products such as organic products, local products, and products supplied with fair trade and by the ability of producers that can meet this demand with their innovations in production. As a result, these agricultural products have become niche products with limited demand and considerable demand uncertainty.

Matching supply and demand is a challenging task in all industries. However, long lead times, inflexibility to change production capacity after initial commitment, effect of uncertainty in supply, and possibility of disruptive shortages make this task even more challenging in agricultural supply chains. Moreover, high demand uncertainty in supply chains for premium fruit and vegetable coupled with high supply uncertainty make it necessary to develop effective planning approaches.

Recent advances in planning methodologies that capture risks in supply chain management combined with the advances in data collection in agriculture allow deployment of these planning approaches in agricultural supply chains.

Although there are many studies and applications of risk-based planning approaches in different industries such as textile apparel industry, consumer goods, etc., the number of studies focusing on agricultural supply chains is limited. For a comprehensive review of studies conducted in the area of production planning problems in agricultural supply chains, the reader is referred to the review of Ahumada and Villalobos (2009). In the literature, most of the studies in agricultural planning focus on crop planning (Romero, 2000; Itoh et al., 2003), harvesting decisions such as how to allocate transportation equipment, scheduling of packing and processing plants, amount of harvest per period for single picking plants such as flowers, wheat, potatoes, etc. (Widodo et al., 2006; Ferrer et al., 2008), and crop choice models (Schilizzi and Kingwell, 1999; Maatman et al., 2002).

Merrill (2007) presents a planning problem for premium tomatoes similar to the planning problem of Syngenta Seeds presented in Jones et al. (2003). Kazaz (2004) considers a production planning problem of a company that produces olive oil in the case of random yield and demand. Allen and Schuster (2004) develop a mathematical model to control the harvest risk in grape farming by scheduling of the harvest. In an earlier study, Doğrusöz et al. (1974) defined a sugar beet planting program to determine the farm areas for beet farming in Turkey and discussed the effects of risk. This problem is similar to the one we discuss in this chapter but it is presented from a central planner point of view rather than from that of an intermediary agribusiness firm working with a number of contracted producers.

In this chapter, we summarize our recent work on developing a risk-based planning methodology for a firm that evaluates different farms for contract farming for premium vegetables and fruits under yield, harvest, maturation, and demand uncertainty.

14.2 Contract Farming

Development of the premium vegetable and fruit customer market allows smaller producers to play a role in the supply chain. The need to supply premium products regularly to major retailers with the desired quality leads these smaller producers to work with larger intermediary agribusiness firms. Consequently, contract farming is used extensively in these agricultural supply chains.

Contract farming has been defined as an agreement between farmers and firms for the production and supply of the agricultural products under forward agreements, frequently at predetermined prices (Eaton and Shephard, 2001).

Under the contract system, the farmer agrees to supply products according to the specifications of the contract in terms of quantity, quality, price, and time. Contract farming helps integration in agricultural supply chains. The use of contract farming improves small farmers' access to the markets, decreases the risk of farmers, and ensures income stability. During the contract term, the contract price can be below or above the spot price at any given time. If the contract price is set correctly, farmers will benefit from reduced variability in their income. The agribusiness firms can offer technological and management assistance to the farmers. Long-term production plans support the production of higher valued crops. Contract farming also provides more reliable production in terms of quality, quantity, and timing for producers (Eaton and Shephard, 2001; da Silva, 2005).

Contract farming has been in existence for many years. First contracts were employed for sugar production in Taiwan after 1885 and for banana production in Central America in the early 1920s. It had become an important part of food and fiber production in Western Europe by the late twentieth century and since the 1930s it has been widely used in vegetable canning industry in North America and in seed industry in Western Europe (Rehber, 2000).

Turkey has 270,000 km² agricultural area, with around four million farms. About 24% of the population worked in agriculture in 2008. The modern food industry in Turkey began with the establishment of the first sugar factory in Afyonkarahisar in 1926. Although considerable progress had been achieved in agriculture with the annual programs in the 1960s and with structural adjustment programs after 1980, the desired productivity has not been achieved in this industry. The share of food supplied by processing is around 20%, in comparison to the 60% share of processed food in the developed countries.

The first contracts in Turkey were used in sugar beet production with the start of food industry. Since the establishment of the first sugar factory, the entire sugar beet production has been done under contracts (Rehber, 2000).

The second major use of contract farming in Turkey is in growing tomatoes. After China and the United States, Turkey is the third biggest tomato producing country in the world; between 35,000 and 40,000 farm families produced about 6.8 million tons of tomato in 2006 and the biggest part of the production was supplied from the Marmara region. Most of the tomato production for the industry is supplied through contract farming.

Some current examples of companies that use contract farming in Turkey are Tukas Food Company, Tat Food Company, and Anadolu Efes Brewing Company. Figure 14.1 shows the companies that employ contract farming and their products.

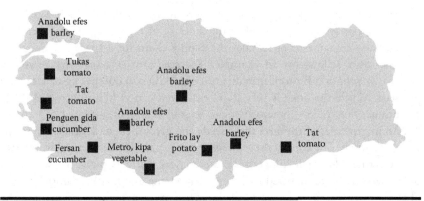

Figure 14.1   Contract farming practices in Turkey.

14.3 Agricultural Supply Chain with Contract Farming

Figure 14.2 shows a simplified agricultural supply chain. In an example of contract farming, an agribusiness firm buys seeds from a seed producer and then contracts a number of farms in different geographical regions to plant these seeds and produce premium fruits and vegetables. Once these products are harvested, the output is transported from all contracted farms to the packaging and processing facilities by the agribusiness firm and then distributed to retailers.

In this supply chain, the intermediary agribusiness firm makes an investment in seeds and then realizes the return from this investment once the seeds are planted, harvested, packaged, distributed, and then sold to the customer. In the premium agricultural supply chain, the firm must manage the risks in the best way possible to match supply and demand.

The seeds for premium fruits and vegetables are sold at different prices depending on the properties of the final fruit or vegetable, resistance to external

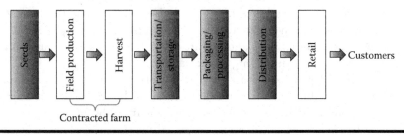

Figure 14.2   Agricultural supply chain.

factors, and their yield. As an example, tomato seeds can be sold at a price of USD 0.01 per seed for commodity tomatoes or at a price of USD 4 per seed for specialty tomatoes. Let us take the supply chain for a specialty tomato—Summer Sun Yellow—as an example. The seeds for this specialty premium tomato are sold by a Biotech firm at a price of USD 350,000 per kg or USD 0.9 per seed. While the price is very high, each seed can yield an output of 20 kg of Summer Sun Yellow tomato that can be sold at a retail price of USD 26 per kg. An intermediary firm buys these seeds and gives them to a number of farms with a contract. The contract sets an agreed price, say USD 5 per kg, to be paid to the farmers for each kilogram of production. Farmers plant the seeds and once these seeds are matured, they start harvesting and continue until the end of the harvest season. Each week, the intermediary firm makes a payment to the farmers based on the total output from that week, transports the quantity that is needed to satisfy the demand that week to a packaging facility, packages them, and distributes to the retailers to meet their demand.

In this environment, not being able to meet the demand is very risky since the company loses the opportunity to make a profit and incurs costs of production, packaging, and distribution. In the same way, producing more than the demand is not desirable either. In order to preserve freshness and quality of premium agricultural products, unused production in one period is not stored for a future period. If the demand is lower than production, the firm leaves the production at the farm in order to avoid unnecessary transportation, packaging, and distribution costs but still pays for the production according to the contract it has with the farms.

The firm must plan its operation so that it matches the demand and supply in the most profitable way. The critical decisions are selection of farms to be contracted, farm areas to be contracted, and the seeding times at each farm so that the demand throughout the year is met by using different farms located in different geographical regions. Especially for firms that work with global customers, selection of farms is a challenging task. For example, in the case of cherries, supplying them to retailers throughout the year requires working with farms located in both Northern and Southern hemispheres at different times.

The seeding time for selected farms is a decision only for annual plants that live only one season such as tomatoes. In this case, the firm must also select the best seeding time so that the plants mature and provide the desired output during the harvest seasons depending on the demand.

For perennial plants such as cherries, the plants live more than 1 year. Therefore, the seeding time is not a decision factor, but rather the farms and areas to be contracted are important decisions that affect the output required to meet the demand.

In a recent study, we developed a stochastic planning method to solve the planning problem of an agribusiness firm that supplies premium fruits and vegetables by using contract farming. The details of the model and the solution methodology can be found in Tan and Çömden (2011) and Çömden (2009). In the next section, the model and the results are summarized.

14.4 Agricultural Planning Problem with Harvest, Maturation, Yield, and Demand Uncertainty

In our model, the objective of the firm is to maximize the total expected profit of a single product that is produced by multiple farms of an agricultural supply chain system over a planning period. The total amount of crop supplied from all farms at any given time is random due to harvest, maturation, and yield uncertainty. The output of each farm depends on the area of seeding and the seeding time of the crops that are the decision variables of the optimization problem.

Historical data that give the maturation time and the length of harvest period depending on the weather conditions allow planners to use the most appropriate distributions for the maturation time and the harvest length during a planning season with the observed weather conditions. Furthermore, historical data on the yield from a farm at a given time is available.

The crops are gathered at the end of each period. Since no inventory can be kept due to the perishability of the product, the crops harvested in one period can only be used to satisfy the demand of the same period. The demand in each period is also random. The companies have detailed historical information about the distribution of demand.

The firm generates revenue from sales of one unit of output to the retailer in a given period. According to the contract, each contracted farm is paid for the whole production. Once the agribusiness firm decides on the farm area to be contracted from a farm, the contract gives the right incentive to the farmer to produce as much as possible from the contracted farm area. The total cost of supplying one unit of production from each farm in a given period includes production, distribution, and packaging costs. If the total supply is greater than the demand in a given period, the firm leaves the excess production at the farms in order not to incur unnecessary distribution and packaging costs.

In each period, the firm generates a sales revenue depending on the total production and the demand in that period. The firm pays for the whole production in this period according to the contract. It also pays for the distribution and packaging costs for the amount supplied to match the demand. The difference between the revenue and the total cost gives the profit for this period. Accordingly, the total profit is the sum of the profits for each period during the planning period. Since the total supply is random due to harvest, maturation, and yield uncertainty, and the demand also is random, the total profit is a random variable. Our optimization problem is maximizing the expected profit by deciding on the best values of the farm areas and also the seeding times for annual plants. Please see Appendix 14.A for the formulation of this problem.

The single period case of this problem is similar to the newsvendor problem with random demand and random supply. By using a general solution to the newsvendor problem when the supply and the demand are normally distributed random variables, we obtained the analytical solution for the single-period single-farm case. This solution involves a nonlinear equation to be solved for the optimal area. We also showed that the optimal seeding time is the one that gives the highest farm availability probability at the period where the demand is realized.

For the single period multiple farm case, we showed that the seeding area decision could be separated from the seeding time decision. More specifically, the maturation and harvest time uncertainty can be mitigated by determining the seeding time while the yield and demand uncertainty can be mitigated by determining the farm area.

For the solution of the multi-period, multi-supplier problem for both perennial and annual plants, we used a normal approximation for the total supply. For annual plants, we derived a set of nonlinear equations that are solved to determine the optimal farm areas. For the perennial plants, we proposed an iterative approach to determine both the optimal farm areas and also the optimal seeding areas. This approximation approach is computationally very efficient and allows us to analyze cases with large number of farms and long planning periods.

Our numerical experiments showed that the proposed approach yields solutions that deviate from the optimal solutions by no more than 1% on average. When the number of farms is large, the optimal solution is not available. Therefore we compared the approximate solution with a mean-value solution where all the random variables are replaced with their mean values. In this case, the resulting problem can be modeled as a mixed integer linear program.

Our methodology that captures harvest, maturation, yield, and demand variability explicitly provides higher profit than the mean-value approach by 15% on average. For perennial plants, the proposed iterative approach that determines both the farm areas and the seeding times yields a profit that is 16% higher on average than the profit obtained by using the mean-value approach. These results suggest that using planning approaches that capture risks associated with harvest, maturation, yield, and demand can be very beneficial for the agricultural industry.

We next discuss a case study in cherry farming in Turkey and discuss the effects of using the proposed solution methodology. In this case study, the company evaluates different farms located in different geographical regions and determines the contracted farm areas to meet the demand during a 4 month period. This selection is currently being done by using a deterministic analysis. By analyzing representative cases, we show that our proposed method, which captures harvest, maturation, yield, and demand uncertainty explicitly to determine the farm areas to be contracted, could increase the expected profit by around 17%.

14.5 Case Study: Alara Agri Business

The roots of cultivated cherry go back to the Black Sea region of Anatolia and extend to the Caspian region and to continental Europe. Around 40% of cherry production is in Europe, and Turkey is the leading cherry producer in the world while the United States is second. Other major cherry producers in the world are Iran, Italy, Russia, Syria, Spain, Ukraine, Romania, and Greece.

Alara Agri Business produces and exports fresh cherries and figs to 22 countries across 5 continents. It was established in 1986 in Bursa, Turkey and now it is the world's largest producer and exporter of fresh cherries and figs.

The company has 750 acres of orchards and 10,000 contracted growers. Throughout the harvest season, every day the products are gathered according to quality specifications. Mobile hydro-coolers are used in order to cool the fruits right after the harvest in each region, and then they are shipped directly to the central pack house in Bursa. The fruits are kept at the ideal temperature through the whole chain. Once the fruits arrive at the packing facility, the quality assurance department checks the product and then the fruits are transferred to the packing line. The fruits are sorted and packed according to the customer demand.

The company supplies cherry to retailers around the world for 8 months—4 months from the farms in Turkey and 4 months from the farms in Argentina. In Turkey, the cherry growing season starts mid-May and continues until early August. In Argentina and Chile, the cherry growing season starts mid-November and continues until mid-January. Figure 14.3 shows cherry growing seasons in different countries of the world.

In Argentina, the company supplies all the fruits from local growers but in Turkey the company produces cherry in its orchards and also buys from the contracted growers. The harvest period of cherry is from May to August in different regions across Turkey. The harvest availabilities can be seen in Figure 14.4. The

	Jan.	Feb.	Mar.	Apr.	May	June	July	Aug.	Sept.	Oct.	Nov.	Dec.
Turkey						▓	▓					
United States							▓					
Italy						▓						
Spain					▓							
Argentina	▓										▓	
Chile												▓

Figure 14.3 Cherry growing seasons in different regions of the world.

	May				June				July				August			
	W1	W2	W3	W4	W1	W2	W3	W4	W1	W2	W3	W4	W1	W2	W3	W4
Manisa																
Izmir																
Bursa																
Sakarya																
Yalova																
Gaziantep																
Mardin																
Çanakkale																
Aydin																
Denizli																
Burdur																
Amasya																
Balikesir																
Tekirdağ																
Tokat																
Isparta																
Afyon																
Uşak																
Çankiri																
Bilecik																
Karaman																
Kütahaya																
K.Maraş																
Konya																
İçel																
Malatya																
Adana																
Niğde																
Kayseri																
Eskişehir																

Figure 14.4   Harvest availabilities of cherry in Turkey.

planning problem for Alara includes the decision of farm areas to be contracted or to be used to satisfy the demand from the retailers during the season.

The company has four cherry orchards, total 300 hectares in Manisa, Çanakkale, Bursa, and Eskişehir. The aim of having orchards in those regions is to assure the cherry supply in early and late periods. Through the main period, the company does not face any difficulties in supplying cherry from the growers that Alara established a contractual agreement with. The forecast about the cherry production is done based on the weather conditions through the year. The weather is compared to the previous years, and the yield estimation is done depending on the year that seems similar to the current year.

Since cherry is a perennial plant, the planning problem for Alara includes only the decision of farm areas to be contracted or to be used to satisfy the demand from the retailers during the season.

A demonstrative problem for this case is analyzed in Çömden (2009) by considering each of the locations given in Figure 14.4 as an alternative farm in the model. The expert opinion of Orchard Operational Director of Alara Agri Business is used to determine the probability function for the harvest and maturation lengths and determine other parameters together with the information in Figure 14.4. For example, consider the farms located in Manisa (see Figure 14.4). The harvest season is expected to start at the beginning of the 3rd week of May and it is expected to last 4 weeks until the end of the 2nd week of June. However, due to the changes in the weather conditions, the harvest season may start at the beginning of the 1st, 2nd, or 4th week of May or at the beginning of the 1st week of June. Moreover, the harvest season may last 3 or 5 weeks instead of 4 weeks. Considering similar changes for all farms located in different geographical regions, this introduces significant variability for the output in each week. In addition to harvest variability, different farms have different yield characteristics. For example, let us assume that one farm is expected to yield 37,500 lb per acre. The realized yield may be above or below this number depending on farm characteristics and external factors. Our observations and discussions show that the yield from a farm can be approximated with a normal random variable with a coefficient of variation that can be as high as 0.3.

Given harvest, yield, and demand variability, determining the farm areas of the contracted farms located in different geographical regions is a challenging task. Without using information about the variability of harvest, yield, and demand, a planner only relies on the expected time harvest starts, the expected time the harvest lasts, and the expected demand to select the farms and to determine the farm areas. Compared to the deterministic solution, the method that captures harvest, maturation, yield, and demand risk directly suggests using different farms located in different geographical regions to reduce the risks. For example, a single farm with an area of 100 acres can be selected to meet the demand by using a deterministic approach. However, when all the risks are taken into account, contracting two farms with areas of 60 and 50 acres and with different costs can be selected. This approach increases the expected profit.

An analysis of different scenarios showed that the planning approach that captures harvest, maturation, yield, and demand uncertainty improves the expected profits by around 17% on the average compared to optimization approaches that only take mean values of random variables into consideration. This improvement is quite significant for agricultural industry.

14.6 Conclusions

Recent advances in agricultural supply chains, especially the growth of the premium fruits and vegetables segment, and emergence of agribusiness firms that work with a number of smaller producers make it necessary to incorporate risk-based planning approaches to match supply and demand more effectively.

Advances in data collection in agricultural industry and advances in weather prediction and recording provide the necessary inputs for planning methodologies that capture supply and demand uncertainty explicitly.

In this chapter, we report our recent research findings on a planning problem of an agribusiness firm that decides on farms and area to contract under harvest, maturation time, yield, and demand uncertainty.

Contract farming is an effective way of managing these risks in an agricultural supply chain. An agribusiness firm that works with large retailers and with a number of contracted farms can pool the supply risk and provide a long term production plan to individual farms and therefore reduce the risks they face. However the firm should match the demand and supply in the most effective way by deciding on the farms, farm areas, and also on when to seed the plants. Our results show that the selection method first selects the farms in different regions to meet the demand in different periods of the year, then decides on the seeding time to mitigate the harvest and maturation time risks and decides on the farm areas to mitigate the yield and demand risks.

Our results show that the proposed approach improves the expected profit by around 16% on the average compared to other optimization approaches that only use the mean values of random variables and therefore do not incorporate risks. Moreover, the benefits increase as the uncertainty increases. As a result, we propose this planning approach as a tool to manage harvest, maturation, yield, and demand risks in agricultural supply chains that operate along the Silk Road.

14.A Appendix

In our model, there are N farms located in different regions. The planning period is T periods. The random production quantity, $Q(t)$, is the total amount of crop supplied from all farms in period t. The area to be seeded in each contracted farm i is a_i. In the planning problem, a_i is a decision variable. Yield denoted with $Y_i(t)$ is defined as the amount of output gained from 1 acre of seeded farm in period t. When the harvest starts at farm i, an output of $a_i Y_i(t)$ is obtained in period t. Historical data on output obtained from each farm gives the distribution of the yield from each farm. We assume that $Y_i(t)$ is a normal random variable.

The period where farm i is seeded is referred as the seeding time and denoted with τ_i, $i = 1,\dots,N$. After the seeds are planted at a given time in farm i, the harvest starts at the completion of a random maturation time. The crops are available to be picked only during the harvest period and the length of harvest period at the farm is also random. Historical data that gives the maturation time and the length of harvest period depending on the weather conditions allow planners to use the most appropriate distributions for the maturation time and the harvest length during a planning season with the observed weather conditions.

The demand in period t is a random variable denoted with $D(t)$. The companies have detailed historical information about the distribution of demand. The firm generates a revenue of $r(t)$ from sales of one unit of output to the retailer in period t. The total production, distribution, and packaging cost of supplying one unit of production from each farm in period t is $c(t)$. The distribution and packaging cost per unit of output is denoted with $s(t)$. We assume $r(t) > c(t) > s(t)$.

The total profit during the planning period can be written as

$$E[\pi(a,\tau)] = \sum_{t=1}^{T} (r(t) - s(t))E[\min(Q(t), D(t))] - (c(t) - s(t))E[Q(t)]. \quad (14.A.1)$$

where

$a = (a_1, a_2, \ldots, a_N)$
$\tau = (\tau_1, \tau_2, \ldots, \tau_N)$
$\pi(a, \tau)$ is the total profit

In the aforementioned equation, the supply at time t, $Q(t)$ is the supply from all the farms at time t. Supply from a given farm at time t is either zero if it is not matured or the harvest period ended, or it is $a_i Y_i(t)$ that is a random variable. As a result, $Q(t)$ is also random. Its moments can be derived by using the probability distributions of yield, harvest time, and maturation time. For example, the expected total supply at time t can be written as

$$E[Q(t)] = \sum_{i=1}^{N} p_{i,\tau_i}(t)E[Y_i(t)]a_i \quad (14.A.2)$$

where $p_{i,\tau_i}(t)$ is defined as the probability that the maturation duration of the farm is completed and an output from farm i is available for harvest in period t given that the seeds are planted in period τ_i. The probability $p_{i,\tau_i}(t)$ can either be derived from the distributions of the maturation time and the harvest time or it can be given explicitly.

The analysis of this problem for the single period-single farm case ($T = 1$, $N = 1$), for the single period-multiple farms case ($T = 1$, $N > 1$), and for the multi period-multiple farms case ($T > 1$, $N > 1$) are given in Çömden (2009) and Tan and Çömden (2011).

References

Ahumada, O. and J.R. Villalobos. 2009. Application of planning models in the supply chain of agricultural products: A review. *European Journal of Operational Research*, 196(1):1–20.

Allen, S.J. and E.W. Schuster. 2004. Controlling the risk for an agricultural harvest, manufacturing. *Manufacturing and Service Operations Management*, 6(3):225–236.

Çömden, N. 2009. Agricultural planning problems with harvest, yield, and demand uncertainty. MS in Industrial Engineering Thesis, Koç University, Istanbul, Turkey.

Da Silva, C.A.B. 2005. The growing role of contract farming in agri food systems development: Drivers, theory and practice. Working document 9: Agricultural Management, Marketing and Finance Service, FAO, Rome.

Doğrusöz, H., İ. Şahin, and M. Parlar. 1974. Türkiye şeker endüstrisi yönetim bilişim sistemi araştırma projesi (In Turkish: Turkish sugar industry management information system research project). Technical Report, Middle East Technical University, Ankara, Turkey.

Eaton, C. and A. Shephard. 2001. Contract farming: Partnership for growth. *Agricultural Services Bulletin* 145, FAO, Rome, Italy.

Ferrer, J.C., A. MacCawley, S. Maturana, S. Toloza, and J. Vera. 2008. An optimization approach for scheduling wine grape harvest operations. *International Journal of Production Economics*, 112(2):985–999.

Itoh, T., H. Ishii, and T. Nanseki. 2003. A model of crop planning under uncertainty in agricultural management. *International Journal of Production Economics*, 81–82:555–558.

Jones, P.C., G. Kegler, T.J. Lowe, and R.D. Traub. 2003. Managing the seed-corn supply chain at Syngenta. *Interfaces*, 33(1):80–90.

Kazaz, B. 2004. Production planning under yield and demand uncertainty with yield-dependent cost and price. *Manufacturing and Service Operations Management*, 6(3):209–224.

Maatman, A., C. Schweigman, A. Ruijs, and M.H. van der Vlerk. 2002. Modeling farmer's response to uncertain rain fall in burkina faso: A stochastic programming approach. *Operations Research*, 50(3):399–414.

Merrill, J.M. 2007. Managing risk in premium fruit and vegetable supply chains. Master of Engineering in Logistics Thesis, Massachusetts Institute of Technology, Cambridge, MA.

Rehber, E. 2000. Vertical coordination in the agro-food industry and contract farming: A comparative study of Turkey and the USA. Research Report, University of Connecticut, Storrs, CT.

Romero, C. 2000. Risk programming for agricultural resource allocation: A multidimensional risk approach. *Annals of Operation Research*, 94(1–4):57–68.

Schilizzi, S.G.M. and R.S. Kingwell. 1999. Effects of climatic and price uncertainty on the value of legume crops in a mediterranean-type environment. *Agricultural Systems*, 60(1):55–69.

Tan, B. and N. Çömden. 2011. Agricultural planning under demand, maturation, harvest, and yield uncertainty. Working Paper, Koç University, Istanbul, Turkey.

Widodo, K.H., H. Nagasawa, K. Morizawa, and M. Ota. 2006. A periodical flowering-harvesting model for delivering agricultural fresh products. *European Journal of Operational Research*, 170(1):24–43.

Chapter 15

Managing Procurement Risks in Turkish Machinery Industry: The Case of Renkler Makina

Muhittin H. Demir, Burcu Adıvar, and Çağrı Haksöz

Contents

15.1 Introduction: Machinery Industry in Turkey

In this chapter, we present the case of Renkler Makina (RM), a major player in Turkish machinery industry, and how it manages its supplier and procurement risks in its global supply chain. Managing risks in the procurement process was of critical

253

importance in the historical Silk Road, as described in Chapter 1. It is getting more important today with the increase in variety of parts, components, and raw materials procured as well as the dramatic extension of the portfolio of suppliers since the old ages. Securing reliable suppliers, managing disruption risks, mitigating price and volume risks always affected the executive mindset while taking decisions.

We begin with an overview of the machinery industry in Turkey. The industry is largely composed of private companies rather than state run companies. This industry is responsible for the livelihoods of nearly 200,000 employees working for around 11,000 companies. These companies are mainly small and medium sized enterprises that have less than 200 employees. About 10 companies are large-scale manufacturers of machine tools and parts, components, and auxiliaries. Manufacturing sites are mainly located in the Marmara, Aegean, and Central Anatolian regions of Turkey. İstanbul, Bursa, İzmir, Konya, Ankara, and Kayseri are the major cities where machine tool, part, and component production takes place.

Overall, the industry is highly versatile, producing machines ranging from building and drilling machinery to hand tools. Being a major market for machinery sales, Turkey took its place among the major producers of machinery (USD 20 billion in 2006). Turkey is listed among the 10 largest markets in Europe for machinery with roughly USD 28 billion sales in 2007. Initially, the main focus of the industry has been on producing machinery for the domestic market. However, since 1980, the focus has shifted toward producing machinery for foreign markets. Table 15.1 and Figure 15.1 demonstrate the export volume changes by the country and export-import changes over time respectively. Main sub-sectors of the industry can be listed as follows:

- Textile machinery
- Agricultural machinery
- Machine tools
- Construction and mining machinery
- Food processing machinery
- Pumps and compressors

We further provide a number of facts regarding the industry in Turkey in order to shed light on the size and the rapid growth*:

- The production volume of the Turkish machinery industry in 2006 was around USD 20 billion.
- The export volume in this sector was USD 8.7 billion in 2007, compared to USD 6.5 billion in 2006.

---

* Available at http://www.us-istanbul.com/pdfs/reports/turkey/turkey_machinery.pdf

Table 15.1 Turkish Machining Industry Export Volumes by Country (in USD)

Country	2006	2007	2008	2009	2008/2009 Rate of Change (%)
Germany	1,059,330,016	1,510,771,879	1,593,734,703	1,100,590,689	−30.9
France	372,225,508	521,777,451	644,391,990	549,185,467	−14.8
England	498,723,164	659,528,108	665,010,648	547,715,577	−17.6
Italy	381,464,078	494,911,696	520,133,753	395,708,582	−23.9
Iran	130,919,982	211,320,147	303,090,318	351,080,135	15.8
Romania	175,077,719	281,099,194	327,173,729	332,640,671	1.7
Iraq	174,437,818	234,570,464	244,814,914	328,719,961	34.3
USA	298,652,098	368,721,569	409,974,222	273,703,555	−33.2
Spain	239,307,101	276,863,649	276,828,222	236,016,820	−14.7
Russia	249,144,109	361,714,999	461,308,212	217,853,164	−52.8
Others	2,937,444,003	3,859,971,508	4,812,129,775	3,797,700,632	−21.1
Total	6,516,725,596	8,781,250,664	10,258,590,486	8,130,406,554	−20.7

Source: Turkish Statistics Institute, Ankara, Turkey, www.tuik.gov.tr

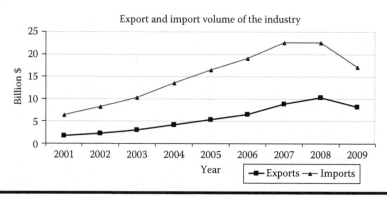

Figure 15.1   Turkish machining industry export and import volumes. (From Turkish Statistics Institute, Ankara, Turkey, www.tuik.gov.tr)

- The import volume was USD 22.5 billion in 2007, compared to USD 19.1 billion in 2006.
- Turkey is among the largest 10 markets of machinery in Europe, with around USD 28 billion total machinery sales.
- Turkey is the 16th largest manufacturer among 29 machine tools manufacturing countries.
- By 2023, industry aims to increase its export volume by 17.8% as shown in Figure 15.2.

The Turkish machinery industry has a wide product range. Today, the industry produces building machinery, heavy industrial machinery, machine tools, hand

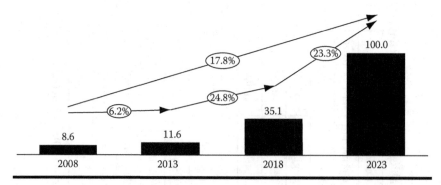

Figure 15.2   Future projections for export volumes (in billion USD). (From 2023 Turkey's Export Strategy Project. Available at www.sanayi.gov.tr/.../makina_sektor_raporu-temm-16082010143342.pdf)

tools, drilling machines, pumps and compressors, textile machinery, food processing machinery, internal combustion engines and turbines, sewing machines, refrigerators, washing machines, dishwashers, valves, conveying and hoisting machines, cutting and bending machines, air conditioning units, woodworking machinery, boilers, and burners. The industry is also capable of producing parts and accessories for the aforementioned machinery types. As of 2009, the share of the local inputs used in the production of these machineries is around 80%.

15.2 Renkler Makina: Company Background

RM is owned by Balaban Group, a family owned venture conducting business in automotive and construction industries. The family started RM in 1980. Today, RM employs more than 260 people and has more than USD 20 million turnover in 2009. As of 2009, the export volume of the company comprises 36% of its annual sales.

The short term strategic target of the RM is to increase its turnover by 20% in current and potential markets. In the long run, the top priority of RM is to meet the requirements of customers beyond expectations, while investing continuously in people and technology. RM lists its strategic goals as follows:

- Serving firms with proved confidence
- Becoming a firm approved by quality
- Becoming the supplier of the five biggest electromechanic companies in the world
- Creating our own brand

RM is one of those firms that experienced a tremendous growth since its establishment in 1980. Part of this growth is due to a truly implemented successful supplier development program of one of their major customers, Schneider Electric. The scope of the program involved the transfer of machinery and equipment for a number of critical punch and press processes from Schneider Electric to RM. While this was a significant opportunity for the growth of RM, it also posed challenges in terms of managing the new processes. Now, RM is an industry leader in metal sheet production and machining besides being a critical supplier for global supply chains such as those of Schneider Electric and AREVA. A recent facility was established in Kemalpaşa, İzmir, Turkey in 2001. The total area of the new facility is $10,300\,m^2$ with $5,000\,m^2$ indoor area. Within the next year, RM has increased its number of employees from 25 to 70 and also started to work with another key customer, AREVA. In addition, RM invested in machine inventory.

The products and services RM provides include iron sheet processing, metal cutting, press, installation processing, welding processing, metal sheet production, and several machining services to electromechanics, automotive, and energy industry. The countries these products are exported to include Germany, China, Russia, France, United Kingdom, Spain, Switzerland, Austria, Belgium, Algeria, and Greece.

Before becoming a key supplier of important global supply chains, RM has gone through serious transitions in terms of its business processes and risk management strategies. A destructive fire in its major warehouse in 2005, substantially disrupted RM operations and brought certain problems to the surface. After a closer look at RM business practices, the following problems were identified by the top management:

■ Irregular supply chain process and lack of procurement strategy
■ Unawareness about the risks exposed pertaining to supplier selection and the procurement processes
■ Inability to measure the performance of the procurement process

Starting from 2005, risk management has become a major issue for RM. Today, top management aims to know what kind of risks the company may encounter in the supplier selection and procurement processes. The managers perceive this as a key component of their overall attempt to improve the business processes and develop an effective supply chain strategy. With the objective of stable growth, RM aims to support institutionalization by focusing on procurement processes through setting standards for selecting the correct suppliers based on scientific methods. Thus, RM will be able to

1. Determine the right procurement strategy
2. Decide on the criteria (including supplier selection, evaluation, and risk management) for performance measurement in the procurement process and relative importance of each criterion (weights) accordingly
3. Take smart decisions by defining the risk-induced actions in the company and the optimum risk levels regarding supply chain and procurement processes
4. Design and implement internal and external risk evaluation surveys that will also constitute infrastructure of supplier evaluation framework
5. Achieve the process integration within a year in order to obtain better purchasing performance evaluation

Regarding the procurement process, there are two important concerns for RM as identified by the top management. The first one is the measurement of performance in order to be aware of the company's procurement process, current and previous practice, identify related inefficiencies, and finally increase the customer service levels. The second concern is the identification and management of risks associated with suppliers. Specifically, RM focuses on reducing the losses due to external risks originating from suppliers.

15.3 Aligning the Company in the Global Supply Chain

RM serves two key customers: Schneider Electric and AREVA. From the supply chain perspective, RM is positioned in a network of supply chains (see Figure 15.3 that illustrates the global supply chain of RM). The focal companies of the supply chain are the customers, Schneider and AREVA, both of which are global players. These companies have high quality expectations, tight schedules, and also demand an increasing share of RM's production capacity. Since the customers require deliveries on exact scheduled dates, RM has to operate in a just-in-time (JIT) environment. Moreover, the competition leaves RM with very low profit margins. Under these conditions, it is inevitable for RM to efficiently manage its supply chain upstream for survival. However, given the fact that customers of RM also hold the power, RM does not always control the decision making initiative to guide the upstream players. Thus, emphasis while managing the supply chain shifts to RM itself and its downstream suppliers.

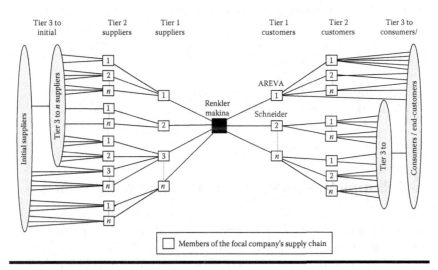

Figure 15.3   Global supply chain of RM.

Acting as a key supplier of global companies on one side, RM works with a large number of suppliers that are of small size (10–20 employees, low capacity) and job-shop type companies. These companies own traditional production equipment and have limited production capacities. They usually dedicate their capacities to single customers. However, in practice, these job-shop type companies cannot work via long term contracts. This is due to several reasons: quality, survival, financial, and structural problems such as type of ownership, conflicts among shareholders, inability to follow technological improvements, lack of engineering know-how, and scarcity of skilled staff. There are approximately 30 companies that are of key importance to RM in this category.

Examining the overall common structure of the key suppliers shapes the downstream supply chain management of RM. It turns out that RM should focus dominantly on risk management techniques in this context. However, it should be noted that, this is different from managing risks of internal operations. It is management of the risks associated with inadequate performance of RM positioned between the suppliers downstream and customers upstream of the supply chain. Symptoms surface with increasing frequencies of not meeting the deadlines and the quality requirements, being unable to react quickly to demand fluctuations, and hence operating with long lead times.

The discussion points to the importance of supplier management process within the context of supply chain integration. Following the goal stated as "institutionalization," RM aims to improve its business processes and develop an effective supply chain strategy. In addition to a variety of implementations such as the "5S" effective workplace organization or warehouse layout improvement, RM aims to redesign the procurement process. However, these attempts seem more like isolated approaches and are far from being the milestones of an overall systematic view. The main goals concerning the procurement process, as stated by the managers of RM, are

> ... setting standards for selecting the correct suppliers and performance evaluation, determining the right procurement strategy, determining the risks associated with supply chain and procurement processes...

15.4 Supplier Management: The Current Practice

Managing the suppliers involves continuous supplier selection and evaluation processes. For these purposes, the current practice at RM uses a composite measure with three criteria and associated weights. The criteria are "quality," "delivery," and "price" with weight factors of 40%, 40%, and 20%, respectively.

Even though there is a statement of policy for supplier management, the policy is incomplete; it lacks a description of the supplier selection process and an assessment of the risks. There is no well-defined structural and operational methodology in terms of information infrastructure, evaluation frequency, or actions to be taken following the evaluation. The purchasing department performs the evaluations, in a rather subjective way and there is no guarantee that the process will be conducted in the same way the next time. The process does not provide an evidence of the overall purchasing performance of RM either. Management of the risks associated with suppliers is left out of the scope of supplier management process. Although the top management has realized the importance of the impacts of the risks, the practice is rather reactive: whenever a risk is realized, a band-aid fix is sought and applied.

15.5 Redesign of the Supplier Selection and Evaluation Process

Supplier evaluation framework relies on the three aforementioned criteria: "quality," "delivery," and "price." Every supplier is scored using the same set of weights (40%, 40%, and 20%, respectively). The adequacy of the criteria is clearly questionable. Factors like the criticality of the material/service purchased, volume and frequency of shipments from a particular supplier, availability of alternate sources for the material/service purchased, previous history of the relationship with supplier call for a framework with different weights for criteria based on the particular material, service, or supplier. One approach would be the identification of groups of suppliers where each group has its own set of criteria and associated set of weights. Since RM is working with a large number of suppliers, with possible inclusions and exclusions from the list over time, the overall process turns out to be complicated. Redesign of the supplier selection and evaluation process with a new set of criteria, along with the identification of groups of suppliers and sets of weights promises a dynamic management of the overall process.

To this end, the first step taken was the selection of the new set of criteria. The criteria need to be well agreed upon and accepted by the management and the owners of the procurement process. Otherwise it is not possible to build a successful system. Critical decisions regarding the new process design have to be made by the decision makers and the implementers. The long list consisted of the well-known criteria listed in Dickson (1966). This list was taken to be a starting point and a basis for discussions. There are 23 criteria in Dickson's framework, namely:

... quality, delivery, previous performance, warranty and compliant policy, manufacturing process and capacity, price, technical capacity, financial position, compliance to procedures, communication system, reputation and market situation, eagerness for job, management and organization, process control, repair and service, attitude, effect, packing skill, relationship, geography, previous jobs, education, and mutual agreement (Dickson, 1966).

The next step in the design of the new process involves a ranking of these criteria based on the judgment of RM managers. This leads to a truncated list of criteria to replace the previous ones. One other aspect of the redesign is the introduction of sub-criteria. For a particular supplier, the score out of one criterion is obtained as the weighted sum of scores for the associated sub-criteria. Benefits of introducing sub-criteria are twofold: First, it provides a more detailed systematic way of computing the scores for each criterion, based on the sub-criteria and the associated weights. Second, it allows for differentiation in the evaluation. RM can simply identify groups of suppliers and the weights for each group, possibly keeping the weights uniform for the main criteria and varying the weights for the sub-criteria. This would result in a methodology that is easier to implement and communicate to suppliers, even to customers and third parties.

The ranking step resulted in six main criteria. These are "quality," "delivery," "price," "relationship," "reputation," and "miscellaneous." The last criterion was proposed to be added into the list by RM managers in order to allow for the future inclusion of possible supplier specific sub-criteria in the evaluation process. However, it was decided to start with an explicit statement of the sub-criteria pertaining to the "miscellaneous" criterion so that implementation is possible. For each of the main criteria, the sub-criteria were defined as follows: The main criterion "quality" is composed of the sub-criteria "possession of quality system certificates," "acceptance rate," and "efficiency of quality system." "Delivery" has sub-criteria "on time delivery," "lead time," "traceability," "convenience of geographical position," and "order fill rate." Under the main criterion "price," we have the sub-criteria "price competitiveness" and "payment options." "Relationship" has "reliability," "service level," "communication," and "mutual agreement" sub-criteria. The criterion "reputation" is based on the sub-criteria "performance history," "professionalism," "references (clients and partners)." Finally "miscellaneous" is measured using scores on "production capacity," "flexibility," and "financial statement."

Once the main and sub-criteria are agreed upon, it is necessary to decide on the weights for each criterion. Thereafter, the system can be used to evaluate each supplier on a uniform basis. However, as discussed earlier, it is desirable to classify suppliers, which makes it possible to evaluate each supplier with respect to the relative significance of criteria as specific to the supplier. Hence, the process requires the identification of the supplier groups, then the weights of criteria for each supplier group. These decisions are viewed primarily as corporate decisions and are mainly taken by RM managers.

Even with a comprehensive statement of the criteria, sub-criteria and the associated weights for each supplier group, the supplier selection and evaluation process is still incomplete. A systematic process requires the definition of the measurement methodology for each sub-criterion. This, in turn includes clear statements and records of the measurement methodology, calculation steps (if any), frequency of measurement, and a definition of how and where the records are to be kept. This emphasizes the utmost importance of the integration of information technology infrastructure of the company into the supplier selection and evaluation process. This integration will enhance the accuracy, validity, and the consistency of the process, hence supplier management.

15.6 Supplier Risk Management

Concerning the procurement process of RM, there is little in terms of managing the risks, and even less on the systematic front. As discussed earlier, the management is aware of the possible outcomes of risk ignorance. Further benefits of effective risk management for RM would be the identification of potential risks in the procurement process, possibility of designing methods to monitor and analyze the factors that may mitigate these risks, and ability to make action plans in order to cope with the situations where the risks are actually realized. As mentioned earlier, risk management in RM is currently perceived and practiced as an operational rather than strategic activity. Moreover, it is primarily reactive. That is, when a risk is realized, a band-aid fix is sought and applied. Typical examples are seen where the contracts with customers include a tight delivery deadline. RM plans for these deadlines based on supplier lead times. In many instances, the delay from a supplier is tried to be covered by rushing the rest of the process carried out in-house by RM, or moving part of the process to an alternate supplier. In some of these cases, RM is unable to prevent the delay and has to offer price discounts or face penalties. If the material under consideration is a common material, RM usually keeps a safety stock in its facilities, formed up by extra units added to the previous orders.

The current methodology used for identification and evaluation of the risks is inspired by a well-known management tool, i.e., SWOT analysis (Weihrich, 1982). The term SWOT emanates from the capital letters of the words *S*trengths, *W*eaknesses, *O*pportunities, and *T*hreats. The tool aims to develop a strategy upon the identification of strengths, weaknesses, opportunities, and threats associated with a company.

The process identifies sub-criteria for each criterion (strengths, weaknesses, opportunities, and threats) with associated weights. The risk score of a supplier is then calculated as the weighted sum of scores it gets for each sub-criterion. This methodology, however, poses problems, both in terms of design and implementation. It is rather straightforward to translate weaknesses and threats into risk scores, but it is conceptually difficult to associate strengths and opportunities with risks.* One alternative would be assigning negative scores to strengths and opportunities, where the process design would require RM to eventually decide on a scale counter effects, for instance, of a possible threat versus a possible strength. Another approach would be assigning positive scores to strengths and opportunities as risk factors. This is conceptually incorrect and again requires the company to decide on a ranking of the effects of a possible threat and a possible strength. In terms of implementation, current methodology is impractical. Consider, for instance, having high production capacity identified as a "strength," and financial instability identified as a "weakness" for a particular supplier of RM. The first alternative would go for assigning a (negative) risk weight to high production capacity and a (positive) risk weight to financial instability. This would have to involve an assessment of relative weights of the two factors; which in other words calls for identifying for any level of financial instability, a corresponding level of production capacity so that the two will "neutralize" each other. With respect to the latter approach, the company needs to assign positive risk weights to both high production capacity and financial instability. The main problem with this is that it does not make sense to assign a positive risk weight to high production capacity that is already identified as a Strength of the supplier. Besides, currently, the criteria are vaguely defined and measured by "Yes/No" type answers, which is in most cases, very difficult to assess.

The first step in the risk management process is to be able to conduct effective risk assessment. This step relies on the urgency of providing a sound identification and measurement of risks for immediate use in managerial decision making. This is specifically critical for RM, primarily due to the aforementioned roles of RM both in the upstream and downstream supply chain.

The suggested methodology for risk assessment was developed based on the FMEA (failure mode and effects analysis) technique. This technique is commonly used in manufacturing facilities for risk assessment. The FMEA implementation requires a site visit in order to determine risk factors. The next step is the identification of estimated frequency of occurrence, level of impact and the likelihood of detection once the risk is realized. The score for each risk is obtained by the product of frequency, impact, and detection values associated with the risk. The method for assessing the risk associated with a supplier then

---

* Besides, risk has the meaning of "opportunity" in Chinese.

may be possible by assigning weights and obtaining the weighted sum of scores of each risk factor, as obtained by the FMEA technique. Aside from a standalone tool, the risk output of this methodology may be integrated into the supplier evaluation process, by defining "risk" as the seventh criterion to the list of main criteria (that included quality, delivery, price, relationship, reputation, and miscellaneous). The implementation is not finished, yet the expected immediate outcome of the process redesign includes a procurement process that is aligned with the business strategy of RM.

Clearly, a successful implementation of such a risk management methodology will not be possible without contribution from the suppliers of RM themselves. This, in turn, requires a correct explanation to an agreement of the suppliers. After all, risk assessment will inevitably lead to significant decisions regarding suppliers, such as purchase volumes, contract prices, and delivery lead times. At this point, the proposed methodology possesses the advantages of being transparent and uniform for each supplier. Moreover, it will be based on actual recorded data and this increases the likelihood of acceptance by any party. However, it would be no surprise that the managers of the procurement process will have to spend a good deal of time and effort in order to correctly convey the methodology to suppliers and receive their acceptance. A good idea could be to employ a joint implementation of the evaluation and follow decisions together with the associated supplier during the transition period.

15.7 Road Ahead

RM is currently undergoing continuous growth due to increasing market share. This necessitates a re-definition of the business environment, based on the supply chain alignment, along with the expansion of their supply chain network. The operational response of the company to this new situation turns out to be the construction of two new warehouses, close to the customer locations, in Turkey and a new manufacturing plant in Algeria.

It is clear that the managerial response of RM needs to be based on future supply chain contingencies and uncertainties. This, coupled with the fact that RM wants to stick to the strategic goal of brand commitment, clearly requires an integral procurement and distribution strategy with a more systematic approach. Inevitably, the current traditional approach for supplier and risk management is no longer practical.

To this end, RM has already started to reconstruct its supplier management processes and recorded certain improvements in terms of supplier selection and evaluation. In terms of risk management, RM redesigned the process of risk evaluations as well as risk management. However, as with a journey that involves continuous improvement, there is still a long road ahead. As the next step, RM managers need to reveal the actions that will be based on the outputs of the supplier evaluations and risk assessment. These may include guidelines for supplier selection, systems for monitoring delivery and/or lead time risks, action plans for unexpected raw material price volatility and raw material availability problems, methodologies for resolving raw material quality issues, and guidelines for communication with suppliers.

Clearly, outputs of the supplier evaluation and risk assessment also form a basis for supplier development programs, including the monitoring of suppliers' capacities, financial and technological health, and alternate suppliers. One other important outcome will be the inclusion of RM's suppliers and customers into these processes, through integrated procurement strategies. This will definitely require more collaboration along the supply chain.

References

Dickson, G. 1966. An analysis of vendor selection systems and decisions. *Journal of Purchasing*, 2:28–41.

Weihrich, H. 1982. The TOWS matrix—A tool for situational analysis. *Long Range Planning*, 15(2):57–62.

Chapter 16

Supply Chain Risk and Sourcing Strategies: Automotive Industry in Iran

Hoda Davarzani and Andreas Norrman

Contents

16.1 Introduction

One important area of research in the field of supply chain management is supply chain risk/uncertainty management, which was given more attention over the last 10 years due to earthquakes, economic crises, SARS, strikes, terrorist attacks, etc. A disrupted supply chain may encounter serious problems in meeting customer demands, it also suffers from a significant drop in supply chain's short-term (Norrman and Jansson, 2004) and long-term performance (Hendricks and Singhal, 2005a,b). An event for a single member of a supply chain can easily influence several members of the chain, causing consequences far beyond the immediate effect at a certain location (Sinha et al., 2004, Bogataj and Bogataj, 2007). Supply chain risk management processes aim to reduce the impacts or likelihood of an event by implementing actions including "avoid, reduce, transfer, share or even take the risk" (Norrman and Lindorth, 2005).

Supply risk is the most important risk of supply chains (Tang, 2006b), and might be the initiative of other risky events as well. There is a fine line between risk and uncertainty concept; risk is the combination of undetectability, likelihood, and potential loss of an event (Tammineedi, 2010) while uncertainty is defined as the absence of information (Brindley and Ritchie, 2005). Risks are mostly determined based on their source (Juttner, 2005), e.g., political risk, economic risk, terrorism, etc. On the contrary, some researchers refer to risks according to the area of their impact (Kleindorfer and Saad, 2005, Tang, 2006b), or even their severity. Supply and demand are the most investigated issues of the supply chains; it is because of their direct impact on financial indicators and overall performance. Among the supply chain risk types are disruptions that result from natural disasters, labor disputes, supplier bankruptcy, and acts of war and terrorism (Chopra and Meindel, 2007). Disruption risks generally have a low probability and the potential for a great loss. Some papers refer to them as "catastrophic events" (Knemeyer et al., 2009). These risks can seriously disrupt or delay material, information, and cash flows, which can spoil sales, increase costs, or both. The supply disruption has more influences on the enterprise performance than the demand disruption, while closer cooperation with suppliers is extremely important to mitigate supply disruption risk and improve performance (Xiaoqiang and Huijiang, 2009). This chapter concentrates on supply disruption and one of the most validated solutions to mitigate its impact, i.e., alternating sourcing strategy. In the following sections, supply chain disruption and sourcing strategies are clarified based on literature. The theoretical aspect of this chapter is followed by a case study in automotive supply chain as one of the major industries in Iran, which is one of the important geographical regions along the historical Silk Road.

16.2 Supply Chain Disruption

Supply chain disruption is a category of risk that delays or even stops final product, material, information, and cash flows; it can ruin sales, increase costs, or both. How a company gets along such threats depends on the type of disruption and the organization's level of preparedness. The problem for managers is to quantify the benefits of various options and choose a good strategy.

Although some researchers worked on cost (Xiao and Qi, 2008) and demand disruptions (Qi et al., 2004, Yu and Qi, 2004), most of the published works in the field of supply chain disruptions are in the area of supply disruptions. Yu and Qi (2004) demonstrated mathematical models for demand disruptions while Qi et al. (2004) examined quantity discount policy when demand disrupts. Xiao and Yu (2006) developed a game theory model to study evolutionarily stable strategies (ESS) of retailers in quantity-setting duopoly situations with homogeneous goods and analyzed the effects of demand and supply disruptions on the retailers' strategies. Xiao et al. (2007) investigated the coordination mechanism of a supply chain with one manufacturer and two competing retailers when the demands are disrupted. Similarly, Xiao and Qi (2008) studied the coordination of a supply chain with one manufacturer and two competing retailers after the production cost of the manufacturer was disrupted.

Some researchers offered to distinguish between supply chain risk and disruption, as the risk is based on a probabilistic event which can be estimated, but disruptions are mostly the result of chance events which cannot be clearly predicted or estimated. Chopra et al. (2007) focused on the importance of decoupling recurrent supply risk from disruption risk and of planning appropriate mitigation strategies. According to this idea, the behavior of supply chains confronting disruptions and the management approach would have some differences with studied risks. Chopra et al. (2007) illustrated how assigning higher supply share to a cheaper but less reliable source would decrease supply risk when the increase of supply risk has the origin in increasing recurrent uncertainty; and on the contrary when the growth of supply risks is due to the increase in disruption probability, sourcing higher amount from reliable but expensive supplier would be effective in mitigating supply risks.

Most of the papers in the field of disruptions can be categorized into two main groups: papers with the aim of clarification of disruption behavior and those describing strategies to manage it. Within the first group, Hendricks and Singhal (2003) showed that supply chain disruptions have negative effects on financial performance measures, as well as on operating income and return on assets. In another work Hendricks and Singhal (2005a) shed light on the effects of supply chain glitches that result in production or shipment delays and estimated their impact on shareholder wealth. Marley (2006) discussed lean management, integrative complexity, and tight coupling, as well as their relationships with disruption effects. Papadakis (2006), based on an empirical analysis, demonstrated the financial implications of supply chain design, particularly on the differences between pull- and push-type designs. Craighead et al. (2007) illustrated the relationship between supply chain structure and disruption severity based on their observations from different case studies.

Researchers approached the disruption management strategies in different ways and by divergent categories. Tomlin (2006) suggested two different groups of strategies, mitigation and contingency, prior to a disruption and discussed the values of these two choices for managing a supply chain disruption. On the other hand, some believe that although risks may differ from disruptions in some ways, both of them are among nondeterministic problems and contain two parts, likelihood and impact. So, in order to manage disruptions as well as risks, companies can investigate on reducing the probability, severity, or both of them for any given disruption. Norrman and Jansson (2004) studied a fire accident at Ericsson Inc.'s sub-supplier and the company's solution for mitigating the likelihood of such events as a proactive plan. Lee and Wolfe (2003) presented strategies to reduce vulnerability to security losses that may cause disruptions. Kleindorfer and Saad (2005) introduced a conceptual framework to estimate and reduce the effects of disruptions. Pochard (2003) discussed an empirical solution based on dual-sourcing to mitigate the likelihood of disruptive events. Tang (2006a) proposed robust strategies for mitigating disruption effects.

All these strategies to manage supply chain disruptions have a critical assumption that the supply chain managers are not aware of the time of disruption occurrences, but experts can estimate vulnerable parts of the chain, consequently they may define some applicable policies. In general, because of the unpredictability and complex effects of disruption, some researchers (Norrman and Jansson, 2004, Knemeyer et al., 2009) choose proactive approaches.

16.3 Sourcing Strategy

Effective sourcing strategy is one of the most important building blocks of successful supply chain management to handle unreliable supply and stochastic demand (Yu et al., 2009). Decision making on sourcing strategy points to different aspects including the global/local sourcing, number of suppliers, the ordering policy, etc. According to Zeng (2000), "sourcing no longer simply refers to getting the materials at desired prices, rather the decision should be incorporated into the buying firms' operating strategies to support or even to improve the firm's competitive advantages." Especially after tragic events during last decade, e.g., 9–11 events and Hurricane Katrina, supply chain specialists have found the criticality of their decision for the company's destiny.

Appropriate use of global sourcing contributes significantly to the performance of supply chains. Although strategic alliance-based global sourcing when highly specific assets are deployed may enhance a firm's competitive advantage (Murray, 2001), it will increase serious risks. Kotabe and Murray (2004) explored potential limitations and negative consequences of outsourcing strategy on a global scale. According to all the pros and cons, decision makers should find out the best applicable solution for their working environment.

From another angle, defining sourcing strategy points at the decisions on supplier portfolio. Based on this definition, researchers have introduced different strategies including sole, single/dual/multiple, and hybrid sourcing. Zeng (2000) reviewed them besides global and network sourcing based on extensive literature and summarized their characteristics. In a deterministic situation, if the supplier capacity is adequate the single sourcing from the supplier with the lowest unit price would be the best decision. But the real environment is accompanied by uncertainty that brings up the new field of research in sourcing era. Although there are some papers in this field regarding the demand uncertainty (Burke et al., 2007), most of the published works concentrate on supply uncertainty.

Haksöz and Seshadri (2007) reviewed the literature of supply chain management in the presence of a spot market from two different aspects, sourcing strategies and procurement contracts. They discussed the supply contract types to reduce supply and price risks and concluded that providing an abandonment option as an incentive to the seller would enhance the value of the supply chain contract. Based on this idea, Haksöz and Kadam (2009) formulated the breach of contract risk from a supply portfolio perspective and developed a risk metric, coined as Supply-at-Risk (SaR). Later, Haksöz and Şimşek (2010) proposed a model of bundled options, i.e., American type contract abandonment and European type price renegotiation that mitigates the breach of contract risk and enhances the supply contract value. Cousins and Lawson (2007) conceptualized and empirically examined buyer-supplier relationships regarding supply sourcing strategies, relationship characteristics, and firm performance. They signified two sourcing strategies (critical and leverage) and argued that a critical sourcing strategy requires collaborative supplier relationships in order to achieve higher relationship and business outcomes, while leverage sourcing strategies have a direct impact on these same performance outcomes.

Wagner et al. (2009) discussed the dependency of different suppliers' default. They—based on empirical data from automotive suppliers—demonstrated how one supplier failure may influence the default of others. Yu et al. (2009) concentrated on the impacts of supply disruption risks with the choice between single and dual sourcing methods and provided guidelines on how to use each method. Xiaoqiang and Huijiang (2009) researched the measures the purchaser should take to maximize his expected income in the presence of disruption. Martínez-de-Albéniz and Simchi-Levi (2009) represented the suppliers' competition for procurement based on game theory where the suppliers are the leaders and the buyer is the follower. Costantino and Pellegrino (2010) proposed a new approach to evaluate the probabilistic benefits of multiple sourcing in managing the supplier default risk.

One of the main criteria to form sourcing strategy is determining whether the buyer would purchase from offshore or near-shore supplier. It may choose one or a combination of them based on the trade-off between imposed cost and responsiveness (Allon and Mieghem, 2010). An issue that has been rarely taken into account is switching cost (Wagner and Friedl, 2007, Davarzani et al., 2010, 2011), when buyer firm wants to shift from one supplier to another or even change its sourcing strategy from single to multiple sourcing or vice versa. The process of decision making is even more critical when the suppliers might be disrupted.

Regarding the number of suppliers, five sourcing strategies can be assumed—sole, single, dual, multiple, and hybrid sourcing. Moreover, these strategies might be considered from global and local point of view. Although single sourcing improves communication due to close buyer-seller relationship and could cause lower cost as a consequence of economies of scale (Zeng, 2000), the uncertainty of a specific buying–selling situation makes dual/multiple sourcing a reasonable strategy. But it is crucial to find out with which level of uncertainty supply chain should shift to dual/multiple sourcing.

While there are numerous papers on single/dual sourcing (Zeng, 2000, Pochard, 2003, Berger et al., 2004, Wagner et al., 2009) to find the proper answer for the best number of suppliers, only few researchers have worked on multiple sourcing (Tullous and Utecht, 1992, Zhou and Fang, 2009) and its mathematical formulations (Martínez-de-Albéniz and Simchi-Levi, 2009). Most of the companies prefer to reduce the number of suppliers in order to decrease the material supplying cost by omitting the unnecessary setup and negotiation costs. Hence the dominant strategies are single and dual sourcing. When the risk of supplier default is high, companies tend to dual source but what if one supplier goes down and the remaining one causes serious problems due to its position as a single source? Problems of monopoly are crucial when because of political instability or high bargaining power of the seller, the buyer firm is compelled to accept special contract conditions to receive the parts. For example, the supplier may include the statement of continuing relationship based on stability of environmental and political issues in the contract, which allows them to renegotiate or terminate the contract in the case of mentioned situations. One possible solution to this problem is to set sourcing strategy on multiple sourcing that causes competition between alternative suppliers and brings down the probability of renegotiation. In addition, this strategy leads to price competition thus preventing price augmentation. Moreover, hybrid sourcing can be another strategy to control supplier relationships and reduce the risk of lack of material in appropriate time. Table 16.1 summarizes the characteristics of each strategy. In addition, all of them will be explored within a case study.

16.4 Automotive Industry in Iran

Automotive industry is the second most active industry of Iran, after its oil and gas industry. In addition, Iran is the largest automaker in the Middle East (Sapco, 2008). In the late 1960s, Iran had the ability to just assemble auto parts. But from the mid-1970s, companies started to design and manufacture automobiles. Although during the 8-year Iran-Iraq war it was not a priority for the country, the company studied in this chapter (ABC) started to export its product to different countries in Asia, Africa, and South America in recent years. It also established production lines in different countries to expand its market and range of products.

16.4.1 Supply Chain Management of ABC Company

The ABC is the most active and prominent supply chain of automotive industry in Iran and in the whole Middle East with more than 60% of local market share. It has established some production sites overseas in addition to its main assembly plant in Iran.

The main focus of this chapter is on its sourcing strategies for the main production line in Iran to reduce supply risks. It is working with about 500 local and 40 international suppliers. The ordering process for local and global suppliers are completely different, as local ones are mostly part of the lean production plan so the ordering process is based on a so called "Kanban system" which is an integration between suppliers and manufacturer and the lead times generally do not exceed 72 h. For foreign parts providers, the order replenishment is done based on the forecast demand as the lead time might even reach up to 2 months. This long lead time makes global supply process really critical.

The ABC company is responsible for managing a six-tier supply chain for local production sites and also local market. The simplified supply chain of ABC company is shown in Figure 16.1. The main production site is very close to Tehran, between the north and the center of the country, while the second one is in the northeast part of the country and the market segmentation is based on the geographical distance from these two sites and their product portfolio. As this

Table 16.1 Sourcing Strategies and Their Characteristics

	Sourcing Strategy	Risks	Benefits	Details
Number of sources	Sole sourcing	Single sourcing risks and, lack of possibility to shift to another supplier in the case of supplier default	Similar as single sourcing	Applicable only when there is just one available supplier
	Single sourcing	Major difficulties in the case of supplier default, lack of bargaining power, necessity of very close cooperation	Improved communication and collaboration, cost reduction, quality improvement	Provides the least cost in deterministic situation
	Dual sourcing	Higher sourcing cost in comparison to single sourcing	Increasing bargaining power, more competition, possibility of recovery if one supplier goes down	Is proposed in the presence of uncertainty and when the supplier capacity is not enough
	Triple/multiple sourcing	Higher sourcing cost in comparison to single sourcing, lack of close relationship with suppliers	Benefits of dual sourcing and, prevention from the risk of monopoly and problems of single sourcing if one supplier goes down	Is proposed in the presence of high uncertainty and when the supplier capacity is not enough
	Hybrid sourcing	Increase in cost	Possibility of recovery if one supplier goes down	Is proposed in the presence of uncertainty and when the buyer does not want to work with alternative supplier in normal situation
Where to supply	Local sourcing	Low flexibility in cost reduction	High responsiveness, low lead time, less transportation cost, no differences in cultural and legal issues	Applicable mostly when responsiveness is the dominant performance indicator versus cost
	Global sourcing	Low responsiveness, higher risk of supply disruption or delay	Access to low cost lands and low salary labor, more competition	Applicable mostly when cost is the dominant performance indicator versus responsiveness

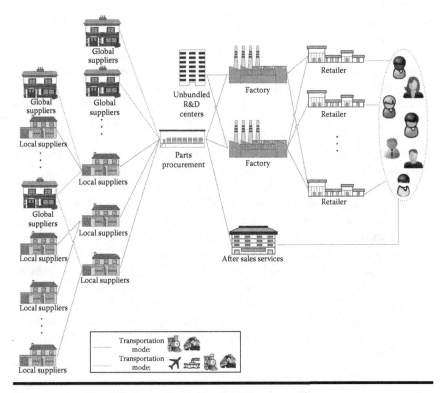

Figure 16.1 Simplified supply chain of ABC company.

company has expanded dramatically in the previous two decades, it unbundled some departments including after-sale services, part procurement department, and engine R&D center.

General policies of foreign suppliers in one specific geographic area are almost similar, so for better supplier relationship management, ABC company divides the suppliers into four categories: East Asia, Southeast Asia, Middle-East and Turkey, and America and Europe. Transportation modes, lead times, and problems of sourcing are similar for the suppliers of each group. Hence, if a supplier in a group goes down, the probability of disruption for its group mates would rise.

16.4.2 Risks Associated with Sourcing in ABC Company

Keeping procurement process in its best condition—regarding the cost and permanency—is the main philosophy of supply chain thought. During recent years, supply chains tried to keep the inventory as low as possible. This phenomenon increased supply chain vulnerability to supply risks and disruptions. All the non-deterministic factors in the sourcing process, including probabilistic lead time and demand, should be considered in the decision-making process in defining sourcing policy. There are several papers, addressing optimal number of suppliers, level of inventory, safety stock, and ordering quantity in association with the distribution function of nondeterministic factors. Regardless of the rich literature for risky events with distribution function, disruptions and their management solutions have not been adequately addressed.

The details of this case study have been extracted from interviews with more than 28 managers and specialist of sourcing in ABC company during 3 years. The supply risk management project started in 2007, and all the risks and their consequences have been identified mostly from interviews and focus group meetings, because there were no comprehensive documents on previous problems and the solutions applied. For proper documentation, a knowledgebase was prepared to record the steps taken, outcomes, obstacles, reports, and forms designed to be used in interviews and meetings. It could provide the supply chain with valuable information so as to facilitate future runs of such a process. Based on the outcomes of data gathering phase, there are five sources of risks for ABC company. Each category reveals specific characteristics of supply chain risks according to its source. Figure 16.2 exhibits these five groups of risks.

16.4.2.1 Organization-Driven Risks

The organization itself might be the source of so many risks. This may rise due to the malfunction of facilities, labor disputes, lack of inventory, technological problems, IT infrastructure troubles, etc. During the study period, one of the activities initiating risk in ABC company was continuous changes in the managerial body and policies of the company. This problem led to serious internal inconsistencies and also troubled relationship with stakeholders, especially in the previous tier of the supply chain. Changes in managerial board were followed by overall changes in most of the company's departments. This was due to lack of commitment to preset organization strategies. This problem could not be solved unless they ran an overall project to align all the company members and departments. In addition, in order to improve the profitability and coordination among supply chain members, the ABC company started a joint venture with some of its suppliers. So, these became permanent suppliers, which led to lower quality as they did not find themselves in competition with potential sellers. But, insisting on a mistake is a bigger mistake, so the company planned to initiate some motivation programs to overcome this problem.

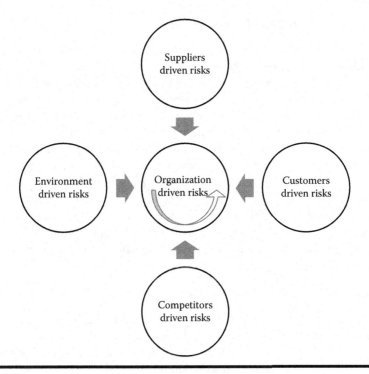

Figure 16.2 Groups of risks in ABC company.

16.4.2.2 Environment-Driven Risks

Based on related literature, environment is one of the main sources of risks for supply chains. The word environment here includes nature, government, global attitudes, society and so on. According to the nature characteristics of Iran as a four season country, if suppliers are not selected smartly, it might cause some sourcing problems in the future. During the last decade, two major natural disasters—the earthquake of Bam and the flood in north Iran—prevented them from supplying materials in proper lead time. Both of the disasters closed the roads and limited accessibility to production sites.

Moreover, one of the main problems of ABC company in recent years has been the unforeseen problems of parts procurement from foreign suppliers. The United Nations and the United States have raised a series of sanctions during last three decades against Iran. Although the automotive industry was not the target in any of them, it has suffered a lot. In addition, fluctuations of exchange and inflation rates have been two of the main issues confronting industries in Iran. Governments have legislated procedures to decrease the impact of these problems, yet they still are concerns of the industry.

16.4.2.3 Competitor-Driven Risks

The automotive industry needs very high initial capital investment, so the risk of new local entrants is very low. Although the majority of local market is served by the company studied, the rivals including public and private local competitors should be considered as an important force. One of the main strategies of the company is to produce a wide range of products including passenger cars and vans to have its share in different market segments. Furthermore, the government supports local industries, which provides more opportunities for their improvement. As a result of high tariffs for importing cars, only a limited range of automobile producers with specific features have found Iran's market attractive. Most of the foreign products with high acceptance rate in Iran have been cars in the medium to high price range that had not been produced in similar class locally. The ABC company took the initiative to approach this market, even though its products were mainly in the low to medium price range segments.

16.4.2.4 Customer-Driven Risks

One of the most cited supply chain risks is the risk of gap between supply and demand (Kleindorfer and Saad, 2005). This may rise due to the differences between customers' request and current designs, and variability of supply or demand. When there are high fluctuations in demand, this risk increases. This phenomenon might be caused by seasonal change or even changes in customers' financial power. Iran is a developing country and during the last decades it has passed through several political–economical crises, so that the financial power of customers has fluctuated in different time periods, and the distribution of demand for different products has changed. Hence, production rate needed to be modified based on more intelligent decisions.

16.4.2.5 Supplier-Driven Risks

In ABC company, suppliers are the main source of risks. The risks observed include, but are not limited to, lack of coordination, inappropriate suppliers' investment, supplier bankruptcy, supplier default (labor dispute, technological problem, etc.), low quality, supplier inflexibility (regarding the new product designs, volume, and lead time), variable lead times and inadequate financial power.

The procurement process is done in relationship with local and global suppliers. According to the previous records of ABC company, local procurement might be mainly disrupted by natural disasters, supplier's bankruptcy and unpredictable problems of supplying from next tier as a result of sanctions.

On the other hand, global sourcing might be interrupted mainly due to sanctions. The aforementioned groups of foreign suppliers, have different positions regarding implementing sanctions and preventing procurement from or shipment to Iran. The ABC company should be aware of possible problems of each supplier based on the records of its group mates. The main problem would arise when a single sourced part cannot be procured due to supplier default caused by any possible event or policy. So the company should think ahead of this problem. One of the main suggested solutions in literature is alternating sourcing strategy based on the level of risk. Although, single sourcing improves communication due to close buyer-seller relationship and could cause lower cost as a consequence of economies of scale (Zeng, 2000), the uncertainty of a specific buying–selling relationship makes multiple and hybrid sourcing a reasonable strategy.

16.4.3 Alternating Sourcing Strategy

Supplier default has been one of the most important problems confronting the supply process of ABC company during recent few years. According to the ideas of managers interviewed, cost reduction and moving through lean supply chain has been the main goal of the company. It has led to a large number of single sourced parts, but in the last 2 years, the sourcing strategy has been switched to dual sourcing and for some cases depending on their specific situations to multiple sourcing. There are five generic solutions to changing sourcing strategy to decrease problems at ABC: Adding alternative local/international suppliers, changing suppliers, revising contracts, adding reserved suppliers, and adding mediators. The main sourcing problems arise when the company is sourcing from one supplier and it goes down or cannot deliver the order. If the company insists on cost reduction, it may prefer to shift to the next low cost supplier but this decision may not be wise enough in nondeterministic circumstances. Consequently, first ABC tried to set the contract statement more carefully, for instance the termination conditions in the contracts were not so clear mentioning just unpredictable events such as wars and natural disasters. So, if European Union or United Nations arranged new sanctions against Iran, the suppliers could provide excuses to disrupt the supply even though the restrictions in sanctions protocols were not targeted for automotive industry. The starting point was to carefully review the contract terms and revise them as required so as to prevent such problems.

Moreover, due to political problems there were some difficulties in parts procurement to Iran. Although it was against the International Air Transport Association (IATA) and international shipping regulations, there had been a few cases where the ship was stopped for further checks after loading in the port or even on international waters. This problem increased lead times enormously and even reduced the production rates. So the company decided to categorize international sourced parts based on the level of risk. The companies in one geographic area had similar backgrounds and were under the same set of governmental regulations. The final groups were categorized based on this factor. This categorization led to the identification of risky parts and the company concentrated on those parts to localize, add alternative supplier, and add mediators to reduce supply risk.

Most of the solutions mentioned have been just temporary cures for the problem, so the company tried to add alternative suppliers both locally and internationally. It was one of the main changes in the company's sourcing strategy; it set dual sourcing as its most preferable policy and wanted related employees in the purchasing department to find out possible alternative sources all over the world. A committee was formed to find out appropriate sources; the least risky events belonged to the East Asian partners of the ABC company so this territory became the first option for source seeking after local potential suppliers. In this way, the company might increase its cost for a period of time, yet the expected cost due to the possibility of supply disruptions would decrease in the future and the risk of production line stoppage could decline even to zero.

But the problem still remained for the parts with low demand as it was not economical to source them from more than one supplier, and sometimes it was even impossible because the suppliers could not accept orders of less than specific amounts. During recent years, the idea of reserve sourcing has been found to be effective in such circumstances. This idea has been mentioned by some researchers as hybrid sourcing, but it is not widely proposed in empirical cases. A company may find a potential supplier for a specific single sourced part, but because of different reasons including conditions of previous contracts or low demand rate, it is impossible to procure that part from both of them at the same time, so the reserve (hybrid) sourcing option would be an acceptable solution. As Figure 16.3 exhibits, assume that part "A" is single sourced and meets the mentioned conditions. Supplier "I" is supplying this part, while supplier "II" is the source for part "B" and also capable of producing part "A." Thus, if supplier

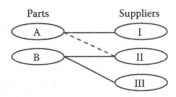

Figure 16.3 Hybrid sourcing.

"I" goes down and cannot deliver part "A," the company can reroute to supplier "B." In order to perform this option, ABC company started potential source seeking for some parts of one pilot product. It means that the presourcing steps including alternating contracts and testing supplier "II" quality for part "A" have been done; and a combination of local and international suppliers are identified to procure pilot parts in the case of disruption. Hybrid sourcing mostly provides the buyer with lower costs than dual and multiple sourcing, and because of relationship with alternative sources, the switching time from Supplier "I" to "II" in the case of disruption decreases.

Previous sections briefly illustrated the supply risks confronting ABC company and how it managed to mitigate them. One of the main influencing issues in managing supply chains is to determine the appropriate sourcing strategy in general and adequate sourcing policy for each part. Figure 16.4 summarizes this progress in ABC company.

Along with the process of determining a suitable sourcing strategy, three main questions should be answered:

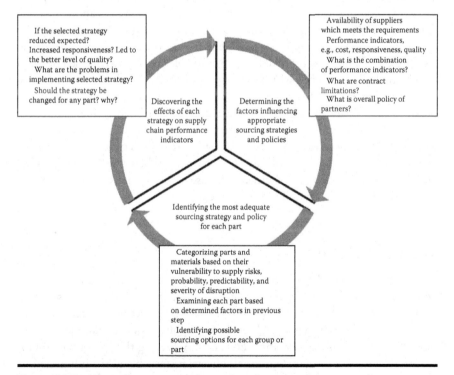

Figure 16.4   Overall process to determine the appropriate sourcing strategy.

1. What are the factors influencing appropriate sourcing strategy?
2. Which strategy works best for each part or group of parts?
3. What is the effect of the selected strategy on supply chain performance indicators?

In addition to the specific requirements of procurement of each part, limitations of standards and performance indicators should be determined. For instance, the company should clarify if it aims to reduce the cost, increase the responsiveness, or even accomplish a particular combination of all indicators; it should identify the most important criteria in decision making and how they should be assessed.

In previous years, the company aimed to reduce the cost that led it to have a high number of single sourced parts. But recent unexpected problems made responsiveness more important than ever. ABC adjusted its goal to have a non-stop production line with the minimum possible cost. In addition, some suppliers insisted on being the only sources and even in the contracts this was mentioned, so dual/multiple options were not feasible for such parts. Moreover, as the main products of ABC company were under the license of a mother company, the suppliers were also required to have the confirmation license to produce the parts. These two conditions limited the potentially available suppliers and it was difficult to find appropriate ones. In the process of modifying sourcing strategy, the company focused on the most vulnerable parts and the ones procured from the risky group of suppliers. It looked for any available suppliers and based on cost-benefit analysis, it checked which strategy (dual or multiple) provided the company with less expected cost and higher responsiveness in the future. During previous years, parallel sourcing was the only option besides single sourcing. But the idea of hybrid sourcing can overcome the problems of so many single sourced parts with no option of adding new sources. This option is under pilot test for parts of one of their products. Potential suppliers have been identified and required tests have been done. In the case of any future supply disruption, ABC company would be well prepared to cope.

16.5 Conclusion

This chapter discussed how an automotive manufacturer in Iran tried to deal with supply risks by modifying its sourcing strategy. The appropriate sourcing strategy should be determined based on the characteristics of required material/parts, special requirements of each potential supplier, and the main purpose of buyer in the terms of performance indicators. Achieving the appropriate sourcing strategy follows similar steps for all buyers; it starts by determining the factors affecting sourcing decisions and terminates by assessing the effectiveness and efficiency of the selected strategy. The result of final assessment may start a new stream of decision making on sourcing where some modifications are required in influencing factors or even general circumstances of business.

In addition to the probability and severity of risks, Tammineedi (2010) suggested the level of predictability as a measure for criticality of risks. Supply disruptions may be originated by environmental phenomenon or supplier defaults. So, the level of predictability varies and even in some cases not only the risk is unpredictable but also its behavior is unknown in advance. In such cases, it is smart to find out the signals before occurrence of a risky event and its possible consequences. Based on the discovered initiation signs, the buyer can predict disruptions and plan ahead to mitigate or recover from it.

The method that ABC company applied to find out the most likely supply disruption was based on the categorization of international suppliers into four groups besides local ones. Each of these groups had specific features according to their previous records of sourcing problems. The most suitable way to categorize suppliers and consequently supplied parts was based on the geographical location of suppliers. Suppliers in one specific area have almost similar attitudes in any particular situation. And from an environmental disaster angle, the analysis of such events according to the geographical location of suppliers is the most rational way to investigate the possibility of such problems.

In conclusion, if a supplier in a certain region has problems in delivery, it is likely that other suppliers in the same area will have similar problems. The material provided in this chapter discussed an example of how an alternating sourcing strategy can be used to reduce the severity of supply risks. Although the focus on responsiveness was the original impetus behind the implementation of dual/multiple/hybrid sourcing within the case company, these sourcing strategies can also be used effectively to reduce the expected future cost in risky environments.

References

Allon, G. and J. A. V. Mieghem. 2010. Global dual sourcing: Tailored base-surge allocation to near- and offshore production. *Management Science*, 56:110–124.

Berger, P. D., A. Gerstenfeld, and A. Z. Zeng. 2004. How many suppliers are best? A decision-analysis approach. *Omega*, 32:9–15.

Bogataj, D. and M. Bogataj. 2007. Measuring the supply chain risk and vulnerability in frequency space. *International Journal of Production Economics*, 108:291–301.

Brindley, C. and B. Ritchie. 2005. Introduction. In Brindley, C. (Ed.) *Supply Chain Risk*. Ashgate, Hampshire, England.

Burke, G. J., J. E. Carrillo, and A. J. Vakharia. 2007. Single versus multiple supplier sourcing strategies. *European Journal of Operational Research*, 182:95–112.

Chopra, S. and P. Meindel. 2007. *Supply Chain Management—Strategy, Planning & Operation.* Pearson Prentice Hall, Upper Saddle River, NJ.

Chopra, S., G. Reinhardt, and U. Mohan. 2007. The importance of decoupling recurrent and disruption risks in a supply chain. *Naval Research Logistics,* 54:544–555.

Costantino, N. and R. Pellegrino. 2010. Choosing between single and multiple sourcing based on supplier default risk: A real options approach. *Journal of Purchasing & Supply Management,* 16:27–40.

Cousins, P. D. and B. Lawson. 2007. Sourcing strategy, supplier relationships and firm performance: An empirical investigation of UK organizations. *British Journal of Management,* 18:123–137.

Craighead, C. W., J. Blackhurst, M. J. Rungtusanatham, and R. B. Handfield. 2007. The severity of supply chain disruptions: Design characteristics and mitigation capabilities. *Decision Science,* 38:131–156.

Davarzani, H., S. H. Zegordi, and A. Norrman. 2010. Dual versus triple sourcing: Decision-making in the presence of supply chain disruption. In *16th International Working Seminar on Production Economics,* Innsbruck, Austria.

Davarzani, H., S. H. Zegordi, and A. Norrman. 2011. Contingent management of supply chain disruption: Effects of dual or triple sourcing. *Scientia-Iranica: Transaction on Industrial Engineering,* 18, in press.

Haksöz, Ç. and A. Kadam. 2009. Supply portfolio risk. *The Journal of Operational Risk,* 4(1):59–77.

Haksöz, Ç. and S. Seshadri. 2007. Supply chain operations in the presence of a spot market: A review with discussion. *Journal of the Operational Research Society,* 58(11):1412–1429.

Haksöz, Ç. and K. D. Şimşek. 2010. Modeling breach of contract risk through bundled options. *The Journal of Operational Risk,* 5(3):3–20.

Hendricks, K. B. and V. R. Singhal. 2003. The effect of supply chain glitches on shareholder wealth. *Journal of Operations Management,* 21:501–522.

Hendricks, K. and V. R. Singhal. 2005a. The effect of supply chain disruptions on long-term shareholder value, profitability, and share price volatility. Research report, Georgia Institute of Technology, Atlanta, GA.

Hendricks, K. B. and V. R. Singhal. 2005b. An empirical analysis of the effect of supply chain disruptions on long-run stock price performance and equity risk of the firm. *Production and Operations Management,* 14:35–52.

Juttner, U. 2005. Supply chain risk management: Understanding the business requirements from a practitioner perspective. *The International Journal of Logistics Management,* 16:120–141.

Kleindorfer, P. R. and G. H. Saad. 2005. Managing disruption risks in supply chain. *Production and Operations Management,* 14:53–68.

Knemeyer, A. M., W. Zinn, and C. Eroglu. 2009. Proactive planning for catastrophic events in supply chains. *Journal of Operations Management,* 27:141–153.

Kotabe, M. and J. Y. Murray. 2004. Global sourcing strategy and sustainable competitive advantage. *Industrial Marketing Management,* 33:7–14.

Lee, H. L. and M. Wolfe. 2003. Supply chain security without tears. *Supply Chain Management Review,* 1:12–20.

Marley, K. A. 2006. Mitigating supply chain disruptions: Essays on lean management, integrative complexity and tight coupling. Master thesis, Ohio State University, Columbus, OH.

Martínez-de-Albéniz, V. and D. Simchi-Levi. 2009. Competition in the supply option market. *Operation Research*, 57:1082–1097.

Murray, J. Y. 2001. Strategic alliance–based global sourcing strategy for competitive advantage: A conceptual framework and research propositions. *Journal of International Marketing*, 9:30–58.

Norrman, A. and U. Jansson. 2004. Ericsson's proactive supply chain risk management approach after a serious sub-supplier accident. *International Journal of Physical Distribution and Logistics Management*, 5:434–456.

Norrman, A. and R. Lindorth. 2005. Categorization of supply chain risk and risk management. In Brindley, C. (Ed.) *Supply Chain Risk*. Ashgate, Hampshire, England.

Papadakis, I. S. 2006. Financial performance of supply chains after disruptions: An event study. *Supply Chain Management: An International Journal*, 11:25–33.

Pochard, S. 2003. Managing risks of supply-chain disruptions: Dual sourcing as a real option. Engineering Systems Division, Massachusetts Institute of Technology, Cambridge, MA.

Qi, X., J. F. Bard, and G. Yu. 2004. Supply chain coordination with demand disruptions. *Omega*, 32:301–312.

Sapco. 2008. *Golden Booklet of Sapco*. Tehran, Iran.

Sinha, P. R., L. E. Whitman, and D. Malzahn. 2004. Methodology to mitigate supplier risk in an aerospace supply chain. *Supply Chain Management: An International Journal*, 9:154–168.

Tammineedi, R. L. 2010. Business continuity management: A standards-based approach. *Information Security Journal: A Global Perspective*, 19:36–50.

Tang, C. 2006a. Robust strategies for mitigating supply chain disruptions. *International Journal of Logistics: Research and Applications*, 9:3–45.

Tang, C. S. 2006b. Perspectives in supply chain risk management. *International Journal of Production Economics*, 103:451–488.

Tomlin, B. 2006. On the value of mitigation and contingency strategies for managing supply chain disruption risks. *Management Science*, 52:639–657.

Tullous, R. and R. L. Utecht. 1992. Multiple or single sourcing? *The Journal of Business & Industrial Marketing*, 7:5–19.

Wagner, S. M., C. Bode, and P. Koziol. 2009. Supplier default dependencies: Empirical evidence from the automotive industry. *European Journal of Operational Research*, 199:150–161.

Wagner, S. M. and G. Friedl. 2007. Supplier switching decisions. *European Journal of Operational Research*, 183:700–717.

Xiao, T. and X. Qi. 2008. Price competition, cost and demand disruptions and coordination of a supply chain with one manufacturer and two competing retailers. *Omega*, 36:741–753.

Xiao, T., X. Qi, and G. Yu. 2007. Coordination of supply chain after demand disruptions when retailers compete. *International Journal of Production Economics*, 109:162–179.

Xiao, T. and G. Yu. 2006. Supply chain disruption management and evolutionarily stable strategies of retailers in the quantity-setting duopoly situation with homogeneous goods. *European Journal of Operational Research*, 173:648–668.

Xiaoqiang, Z. and F. Huijiang. 2009. Response to the supply chain disruptions with multiple sourcing. In *International Conference on Automation and Logistics*, Shenyang, China.

Yu, G. and X. Qi. 2004. *Disruption Management: Frameworks, Models and Applications*. World Scientific, Singapore.

Yu, H., A. Z. Zeng, and L. Zhao. 2009. Single or dual sourcing: Decision-making in the presence of supply chain disruption risks. *Omega*, 37:788–800.

Zeng, A. Z. 2000. A synthetic study of sourcing strategies. *Industrial Management & Data Systems*, 100:219–226.

Zhou, X. and H. Fang. 2009. Response to the supply chain disruptions with multiple sourcing. In *IEEE International Conference on Automation and Logistics*, Shenyang, China.

Index

Printed in the United States
by Baker & Taylor Publisher Services